Symposium on

ENERGY DISPERSION X-RAY ANALYSIS: X-RAY AND ELECTRON PROBE ANALYSIS, Toronto, 1970

J. C. Russ, coordinator

A symposium
presented at the
Seventy-third Annual Meeting
AMERICAN SOCIETY FOR
TESTING AND MATERIALS
Toronto, Ont., Canada, 21–26 June 1970

ASTM SPECIAL TECHNICAL PUBLICATION 485
04-485000-39

List price $20.00

AMERICAN SOCIETY FOR TESTING AND MATERIALS
1916 Race Street, Philadelphia, Pa. 19103

© BY AMERICAN SOCIETY FOR TESTING AND MATERIALS 1971
Library of Congress Catalog Card Number: 72-137455
ISBN O-8031-0070-1

NOTE

The Society is not responsible, as a body,
for the statements and opinions
advanced in this publication.

Printed in Baltimore, Md.
May 1971

Foreword

The Symposium on Energy Dispersion X-ray Analysis: X-ray and Electron Probe Analysis was given at the Seventy-third Annual Meeting of the American Society for Testing and Materials held in Toronto, Ont., Canada, 21-26 June 1970. The Sponsor of this symposium was ASTM Committee E-4 on Metallography. J. C. Russ, Jeolco (USA), Inc., presided as symposium chairman.

Related
ASTM Publications

Contents

Introduction

At the previous ASTM Symposium on X-ray and Electron Probe Analysis (STP 349) in 1963 only token mention was made of the possibility of energy dispersion X-ray analysis. The development of the wavelength dispersion crystal diffraction spectrometer and of special crystals for it has continued over the past several decades to the point where very highly refined instruments are commercially available. The history of the energy dispersion X-ray spectrometer is much more recent.

The first application of this device to the electron microprobe was reported by two of the authors of papers in this volume (Fitzgerald and Heinrich) in 1968 and the first application to the scanning electron microscope by another author (Russ) shortly thereafter. At this time the performance of the energy dispersion X-ray analyzer was barely adequate to separate adjacent heavy elements and to identify elements down to about atomic number 15.

The companies manufacturing semiconductor detectors and electronics and the national laboratories have made great strides in development within the past few years. The past concentration on application of these instruments in nuclear physics have given way with the realization that a substantial market is available in X-ray analysis for industry and research. Development of systems optimized for the particular requirements of X-ray analysis has proceeded very rapidly; at its present state of art the energy dispersion X-ray analyzer is competitive in many respects with the older wavelength dispersion spectrometer. It is likely that the energy dispersion X-ray analyzer will replace the older wavelength dispersion spectrometer in many, perhaps most, of its applications within the next few years. The papers in this volume discuss, in detail, the relative merits and disadvantages of the two techniques.

Because of the rapid developments in this field there has been no comprehensive literature available. Accordingly, Subcommittee 15 on Microprobe Analysis of ASTM Committee E-4 on Metallography has sponsored the symposium represented by the papers in this volume. The intention was to provide a description of the internal workings of these devices so that the user could understand the design choices that had been made and their importance for X-ray analysis, and to summarize the capabilities and areas of applications of the instruments as a guide to new workers in the field.

One of the first problems the committee had was deciding on the name of the symposium. These detectors have been variously called nondispersive, nondiffractive, solid state or semiconductor, and energy dispersion X-ray analyzers. We rejected the first one because it is both technically incorrect and conveys no information and the second because it did not describe to the user the workings of the device. The third and fourth conveyed meaning only to the manufacturers, not users. The last is technically acceptable and moreover identifies the significant variable by which the X-rays are classified. Therefore we have chosen the name Energy Dispersion X-ray Analysis and hope that the publication of this book with that title will help to simplify and standardize the terminology in the future.

The meeting held in Toronto in June of 1970 was attended by over 200 people and included twelve of the papers in this volume. The papers were invited from the outstanding workers in the field to cover specific topics and present a complete picture of the equipment, capabilities, and techniques of energy dispersion X-ray analysis, especially in comparison to the older, better known wavelength dispersion spectrometers. The first six papers discuss the design of various parts of the system and the ways in which design choices influence results. The last seven papers describe the particular applications of these analyzers to various instruments and the interpretation of the resulting data. All of the papers were reviewed, and one additional paper on light element analysis was written after the symposium, to take advantage of the extensive discussion that took place at the meeting between the attendees and the authors. It is hoped that this volume will become an important reference volume in a very exciting and rapidly growing field.

J. C. Russ

Jeolco (USA), Inc.,
  Medford, Mass. 02155;
  symposium chairman.

Ray Fitzgerald[1] and Peter Gantzel[2]

X-ray Energy Spectrometry in the 0.1 to 10 Å Range

REFERENCE: Fitzgerald, Ray and Gantzel, Peter, "**X-ray Energy Spectrometry in the 0.1 to 10 Å Range,**" *Energy Dispersion X-ray Analysis: X-ray and Electron Probe Analysis, ASTM STP 485,* American Society for Testing and Materials, 1971, pp. 3–35.

ABSTRACT: Energy spectrometry utilizing a solid-state detector (silicon) for X-ray elemental analysis is described, and compared with wavelength spectrometry.

Parameters which influence the spectrometers energy resolution, energy detection range, and countrate capabilities are reviewed. Measured performance of a spectrometer (200 eV FWHM resolution for 6.404 keV radiation) is demonstrated for both X-ray and electron specimen excitation.

KEY WORDS: spectroscopy, spectrometers, X-ray spectrometers, chemical analysis, X-ray spectra, X-ray analysis, electron probes, energy bands, wavelength, X-ray fluorescence, solid state counters, silicon, dispersing

Rapid detection and immediate visual display of X-ray spectra provide a powerful tool for solving many types of analytical problems. Recent advances in the last two years of solid-state detector, preamplifier, and multichannel analyzer technology made it possible to design an energy spectrometer (ES) adequate for X-ray chemical analysis by simultaneous detection, resolution, and display of characteristic spectra from 0.1 to 10 Å. The ES is characterized by: (1) simultaneous detection and recording a major portion of the X-ray spectrum, allowing in one integration period the determination of peak and background intensities for most major elements; (2) rapid visual display of the spectrum; (3) high collection efficiency; (4) less sensitivity to X-ray source position than wavelength spectrometers; (5) sensitivity to a larger X-ray energy range (1 to 50 keV); (6) variable size and shape of detector; (7) no moving parts or mechanical alignments; and (8) lack of higher order lines which are generated in crystal diffraction. These characteristics, coupled with currently available energy resolution, have created new X-ray capabilities, such as rapid visual

[1] Lecturer, Applied Physics and Information Science, University of California, San Diego, Calif.

[2] Chemist, Gulf General Atomic Incorporated, San Diego, Calif.

3

observation of an X-ray spectrum on an oscilloscope screen. The X-ray spectrum characteristic lines and background are recorded simultaneously by sequential sampling of each intercepted X-ray photon, regardless of its energy. In contrast, the wavelength spectrometer can sample only a narrow band of photon energies at one time, and geometrical constraints on the crystal, specimen, and detector positions limit its collection efficiency. Orders of magnitude increase in intensity are obtained readily with the ES due to the short specimen to detector distance that is limited only by the mechanical bulk of the detector assembly.

Diffraction and focusing conditions of the wavelength spectrometer require precise mechanical motions of the crystal and detector with respect to the X-ray source to be analyzed. Thus, the specimen must be accurately located at a particular position in space, lest any deviation cause a loss of intensity. This is not the situation with ES where neither diffraction nor focusing are required. Another advantage of ES lies in its response to a wide range of X-ray energies whose coverage by wavelength spectrometry requires a multiplicity of crystals and detectors.

In spite of all the advantages offered by the ES, it should be remembered that wavelength spectrometry still provides the highest resolution for X-ray energies less than 20 keV. If errors due to counting statistics are small, wavelength spectrometry affords higher accuracy in quantitative analysis and better limits of detectability due to the reduced line overlap and higher peak to background intensities.

Energy Spectrometer System Description

The basic components of a solid-state ES are illustrated in Fig. 1. The solid-state (Si) detector and first stage of preamplification are mounted in a vacuum cryostat cooled by liquid nitrogen. Dynamically pumped vacuum systems of the type presently used in electron microscopes and microprobes contain enough oils to lower resolution by increasing detector surface leakage [1],[3] and degrade detection efficiency by increasing the effective dead layer. To prevent contamination of the detector surface from condensation of impurities, the cryostat's X-ray port is sealed by a beryllium window. A high voltage detector bias supply, preamplifier, band pass linear amplifier with d-c restoration, and a multichannel analyzer complete the system.

Solid-State Detector

The solid-state detector is similar to a gas ionization chamber, in that it is characterized by having no internal gain. The replacement of a gas by a solid increases the number of free charge carriers created by a given photon energy loss. The increased number of free charge carriers, about eight times in the case of silicon, should reduce the relative error from statistical fluc-

[3] The italic numbers in brackets refer to the list of references appended to this paper.

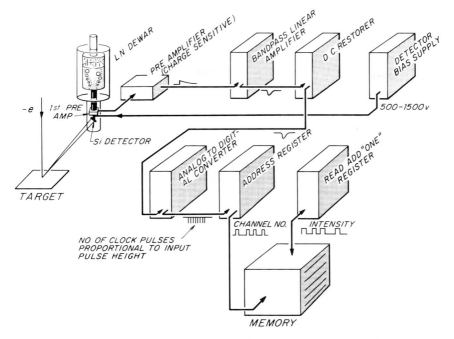

FIG. 1—*Components of solid-state energy dispersion spectrometer (electron excitation illustrated).*

tuations. Assuming for a moment that Poisson statistics apply in the charge generation and collection mechanisms, a comparison of detector resolution indicates an increase of statistical accuracy of the solid over the gas by a factor of approximately three. For example, the resolution of an argon and a silicon detector for copper $K\alpha$-radiation would be:

$$\% \text{ resolution} = \frac{\Delta E(\text{FWHM})}{E(\text{peak})} \times 100$$

$$= \frac{2.35\sqrt{E\epsilon}}{E} \times 100$$

where:

E = energy of detected radiation (copper = 8.04 keV);
ϵ = average energy required to produce a charge carrier pair;
ϵ_{argon} = 26.3 eV/e^-, ion;
$\epsilon_{\text{silicon}}$ = 3.6 eV/e^-, hole, 300 K; and
$\epsilon_{\text{silicon}}$ = 3.8 eV/e^-, hole, 77 K.
The constant 2.35 relates one standard deviation (σ) to the full width at half maximum (FWHM) of a Gaussian distribution.

Thus, the resolution of argon and silicon detectors should be 13.4 and 5.1 percent, respectively, for copper $K\alpha$-radiation.

X-rays (less than 1 MeV) interact with a solid by one of three processes: the photoelectric effect, Compton scattering, and elastic scattering. For X-ray energies below 50 keV, the initial interaction is usually a photo-electric absorption followed by secondary ionization. The energetic secondary electrons from the photoelectric absorption lose their energy very rapidly by two competing processes: impact ionization producing electron-hole pairs and creation of optical phonons. These processes continue until the secondary electrons have insufficient kinetic energy to produce an electron-hole pair. The average energy for electron-hole production (ϵ) is then the sum of the ionization threshold energy, the energy consumed by optical phonons and the residual kinetic energy [2]. Therefore, the average energy converted in the solid for electron-hole pair production must be larger than the band gap energy (E_g). For silicon E_g is 1.1 eV while ϵ is 3.6 eV at 300 K [2].

Because of these competing processes, the total number of charge carriers $N = E/\epsilon$ is reduced and is subject to statistical fluctuations. Contrary to our previous assumption in comparing gas and solid detectors, actual fluctuations are a small fraction of that predicted by Poisson statistics of N independent events. Fano [3] considered the degree of correlation in successive ionizations, introduced a factor F, and derived the variance NF. The expression for full width at half maximum then becomes:

$$\text{FWHM (volts)} = 2.35\sqrt{E\epsilon F} \dots\dots\dots\dots\dots(1)$$

Measurements of FWHM as a function of E indicate an upper limit for F of 0.13 [4,5].

Both gas and solid detectors rely on a high electric field across a region of low conductivity. Neither silicon nor germanium, even in the purest state, has a sufficiently low intrinsic conductivity to prevent leakage current noise from dominating the detected photon signal. This problem is circumvented partly by lowering the temperature of the material. In addition two ingenious schemes are used to increase the electric field and active volume of the detector. They are, respectively, the semiconductor junction (McKay [6]) and lithium compensation (Pell [7]). The semiconductor junction with reversed bias produces a region of high field depleted of free charge carriers, which widens under increasing voltage. Lithium compensation permits further widening of this high field region by neutralization of electrical activities that exist even in the best available materials.

The choice of detector materials is limited by the following requirements:

1. For high energy resolution it must have a low value for the mean energy of electron-hole pair production (ϵ).

2. Carrier life times must be long in proportion to charge transit times to give efficient charge collection.

3. Freedom from carrier trapping centers.

4. At the operating temperature the intrinsic carrier density should be low, since this is related to the problem of current noise.

Liquid nitrogen cooling is necessary not only to reduce thermally created noise in the detector and first field effect transistor, but to prevent lithium migration in silicon and germanium. Days of room-temperature storage of lithium compensated silicon may produce small changes in detector characteristics, whereas germanium is affected severely in 30 min.

The free charge carriers (electron and hole) generated by an X-ray photon, move towards the collection electrodes under the influence of an applied electric field ($\sim$300 V/mm). The induced output from the detector is the result of electron and hole motion, whose relative contribution depends on where in the active volume they originated. The net result is the same regardless of the type of carrier motion. In practice, a large bias is employed resulting in carrier velocity saturation (10^{+7}cm/s). This and the lack of internal detector gain makes the collected charge relatively insensitive to variations in bias potential. As bias potential is increased detector resolution improves due to increased charge collection efficiency. However, at some point, detector leakage current dominates, resulting in a loss of resolution. Figure 2 schematically represents the solid-state detector, biasing technique, and signal amplifier.

Preamplification

A cooled field effect transistor, mounted in close proximity to the detector, in conjunction with several additional gain stages outside of the cryostat forms the preamplifier system. A modern preamplifier is of the charge-sensitive type (Fig. 2), effectively acting as a current integrator[4] and providing an output voltage proportional to the collected charge from the detector.

Unlike the amplification which occurs in the gas proportional detector (gas gain), only the original amount of free charge is collected at the electrodes of the solid-state detector. Noise contributions from the preamplifier are significant and a limiting factor on resolution of low energy X-rays (<10 keV). The electrical noise modulates the statistically varying detector signal (Eq 1) causing an additional apparent energy spread.

The electrical and mechanical design of the preamplifier and the inosculation of the detector to the first amplifier are critical to the spectrometer's resolution, gain stability, linearity, and count-rate capability. The decisive parameter which permits the solid-state detector to function in the low energy X-ray region is the low noise performance of the preamplifier, which directly sets the resolution and lower energy detection limit. Low

---

[4] The charge, or the integral of the total current, is proportional to the energy loss in the detector; the instanteous current may not be proportional.

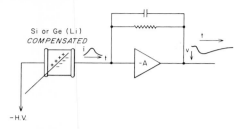

FIG. 2—*Solid-state detector, biasing, and amplification.*

noise is a result of using field effect transistors with the best signal-to-noise ratios at low temperatures [8,9]. Other considerations such as detector capacitance, circuit capacitance, and resistance also play a major role in the preamplifier signal-to-noise performance.

The X-ray analyst also must be cognizant of the fact that mechanical vibration of the detector assembly by external equipment, such as vacuum pumps, or even the boiling of the liquid nitrogen can produce low frequency noise (microphonics [10]). Because of its high gain, low noise characteristics, the preamplifier must be grounded properly and isolated electrically from other systems such as transmission electron microscope (TEM), scanning electron microscope (SEM), or microprobes to prevent low frequency noise from ground loops.

Main Amplifier and D-C Restoration

The capabilities of the semiconductor detector are realized fully only by using a specialized amplifier which is designed to suit the characteristics of the detector and preamplifier.

The functions of the main amplifier are to increase the preamplifier output signal to voltage levels compatible with analog to digital conversion, and provide band-pass characteristics which enable the signal to be extracted with the least amount of noise.

Important main amplifier characteristics are:

1. *Amplifier Linearity*—to preserve linear energy calibration of the pulse height analyzer.

2. *Gain Stability*—to maintain calibration of the energy scale of the multichannel analyzer.

3. *Low Noise*—the main amplifier contribution to the total spectrometer's electronic noise should be negligible.

4. *Rapid Overload Recovery*—overload pulses should return to the amplifier base line in short periods of time.

Most parameters are set by the electronic manufacturers of nuclear equipment, leaving a few for the X-ray analyst such as gain, shaping time, pole-zero adjustment, and the degree of d-c restoration and d-c output level.

Shown in Fig. 3 are the main functions of the band-pass linear amplifier. The gain of the amplifier is varied by an attenuator control which is adjusted so that pulses of interest fall within the dynamic range of the analog to digital converter of the pulse height analyzer. This is accomplished in modern amplifiers by varying the parameters of feedback networks, thereby maintaining a high ratio of preamplifier to main amplifier noise and gain compensation for changes in time constants.

Differentiating and integrating circuits form the band-pass characteristics of the amplifier and are controlled by the shaping time adjustments. The band-pass characteristics of the main amplifier form a window in the frequency spectrum; if the window is of the proper shape it will discriminate against noise. The shape, width, and position of this frequency window determine the portion of the original detected signal and noise which will be sampled and in what ratio. In theory, the optimum output pulse would be cusp-shaped but in practice, for a variety of reasons, it is better to approach a Gaussian shape. While short shaping times give higher counting rate capability, the field-effect transistor (FET) noise spectrum usually dictates an optimum setting for best resolution.

The pole-zero cancellation adjustment corrects undershoot voltage associated with differentiating circuits, reducing one source of pulse height distortion at high counting rates.

Electronic d-c restoration is accomplished by clamping the baseline to some stable reference voltage, and is maintained by d-c coupling to the analog-to-digital converter of the multichannel analyzer.

Multichannel Pulse Height Analyzer

The digitizing, storage, and display of the energy spectrum is done by a multichannel pulse height analyzer.

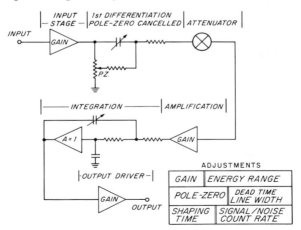

FIG. 3—*Main functions of band-pass linear amplifier.*

The pulse height measuring circuit most commonly used is the Wilkinson type [11]. The analog-to-digital conversion is accomplished by charging a small capacitor to the peak voltage of the input pulse and then discharging it at a constant rate. During the period of capacitor discharge, an address register counts pulses derived from a stable oscillator, and this register subsequently designates which memory address is to be incremented. Every detected pulse accepted by the analyzer adds "one" to a particular memory location during the integration period and the stored spectrum that results may be viewed on an oscilloscope display. This display is operated by re-peated rapid digital-to-analog conversion of each memory location thereby generating a Y-deflection for content and X-deflection for channel. To minimize the acquisition time the digitizing rate of the analog/digital con-verter should be fast (>50 MHz), and the memory cycle should be short (<5 µs). This processing time, which results in lost counts, is com-pensated electronically by extending the integration period an appropriate amount. For quantitative analysis the precision of this correction should be verified by the X-ray analyst.

The calibration of an analyzer's energy scale is accomplished by obtain-ing line positions from known elements. Adjustment of the main amplifier gain determines the energy span recorded by the analyzer.

The measurement of resolution and energy calibration is demonstrated in Fig. 4 using manganese and iron $K\alpha$ characteristic lines. Energy cali-bration, in volts per channel, is obtained by dividing the known energy that separates the two lines by the number of channels that separate the two peaks.

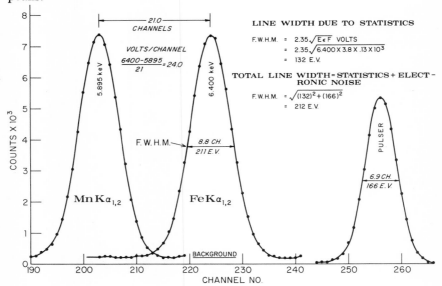

FIG. 4—*Measurement of resolution and energy calibration.*

The electronic noise (detector leakage, preamplifier noise) contribution to the spectral line width is determined by measuring the width of a line generated by a noise free electronic pulse generator connected to the preamplifier input. Generator pulses suffer the same modulation effects from electronic noise as pulses generated in the detector itself. Therefore, the total broadening is the quadrature addition of detector leakage noise, preamplifier noise, and detector statistics (fluctuations in the process of charge production), and incomplete charge collection.

$$FWHM \text{ (volts)} = [(\text{detector fluctuations})^2 + (\text{preamp noise})^2 +$$
$$(\text{detector leakage noise})^2]^{1/2}. \ldots(2)$$

The ratio of spectrum line intensities of standard and specimen to be analyzed, form the basis of quantitative elemental analysis. Three methods for measurement of line intensities are:

1. *Line Height*—height of spectrum line above background. Used exclusively for obtaining rapid intensities from oscilloscope and *X-Y* recorder output. This measurement assumes a constant line distribution shape from specimen to standard and is therefore subject to error when spectrum distortion exists due to high count rate. This measurement exhibits the highest line/background, least error from overlapping adjacent lines, and highest statistical error.

2. *Integrated Line*—the total number of counts (detected photons) contained in a spectral line. Where background is insignificant this measurement has the least statistical error, but the largest line overlap error. The data require channel summing by a multichannel analyzer or computer. Inherently, this measurement is insensitive to shape changes due to spectrum distortion, assuming the summing procedure recognizes new limits of integration when the distribution shape and position has been altered. In this regard, a single channel pulse height selector can perform the integration when the number of elements to be analyzed is known and limited.

3. *±1.4σ Integrated Line (Gaussian)*—the total number of counts contained in the line from -1.4σ to $+1.4\sigma$. The σ limit is derived from the detectability limit[5] assuming a Gaussian distribution. This measurement is statistically more accurate than the line height measurement, but errors due to line overlap are larger. If this criterion is used to set the window of a single channel analyzer, large errors may result from small gain shift or changes in peak shape.

---

[5] Minimum detectability limit fraction

$$C \propto \sqrt{B/I^2}$$

where:

      C = ratio of smallest significant increment that can be detected out of a total, and
I and B = total number of independent events in line and background, respectively.

Pulse pileup and baseline shifts are major distortions which affect the quality of all detected spectral lines at high counting rates. These distortions are derived from the amplifying, pulse height measuring and recording electronics, and result from necessary design compromises for obtaining high resolution, rather than a lack of precision electronics. Obtaining high resolution is the most pertinent factor in eliminating line overlap, but a spectrometer, to be useful as an analytical tool, also must have a large dynamic count rate range. At low counting rates the consideration is one of signal to noise. As count rate demands on the spectrometer increase, the loss of resolution by pulse pileup increases and becomes a dominating effect. Pulse pileup results from pulses stacking on preceding pulses before the amplifying electronics have had sufficient time to settle back to a quiescent state. This introduces an offset voltage causing error in the pulse height measurement, resulting in spread of the spectral line.

The choice of main amplifier shaping times to obtain best energy resolution at low counting rates is usually unique; the shaping times are long compared to what the X-ray analyst encounters in gas and scintillation detectors.

Figure 5 illustrates line broadening as a function of count rate for different amplifier shaping times. The data were obtained from an experimental preamplifier, experimental main amplifier, and a multichannel analyzer (4 MHz digitizing rate), and do not represent presently available systems with regard to high resolution at high counting rates. The counting rate includes iron $K\alpha$ and background produced by 20 kV electron excitation. The multichannel analyzer dead time was 80 percent for 60,000 counts per second

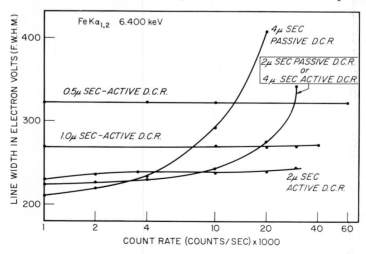

FIG. 5—*Line width as a function of count rate for different main amplifier time constants and modes of d-c restoration.*

(cps) with main amplifier shaping times of 0.5 μs and active d-c restoration. The figure amply demonstrates the effect of increasing line width (FWHM) for longer shaping times (4 and 2 μs) for increasing count rate. Also, the effect of two types of d-c restoration (DCR) are shown. Passive DCR affords superior low frequency noise rejection allowing higher resolution at modest count rates ($<$ 3500 cps as illustrated).

Clearly, short shaping times are more desirable when the X-ray analyst anticipates a large dynamic count rate range; the price paid is loss of resolution at low counting rates. Another approach is to maintain long shaping times and employ a pulse pileup rejection system. However, a correction must be applied to the intensity ratio between standard and unknown to account for the intensity loss due to rejected pulses. A more exacting measurement which is more sensitive to count rate-induced spectral distortions is FWTM. Line position and skirt width (FWTM) are important to the X-ray analyst when a low intensity line is adjacent to a line of high intensity.

Silicon Spectrometer Characteristics

The detection efficiency, energy resolution, and method of specimen excitation determine the ES capability for chemical analysis. Ideally, the spectrometer should have sufficient energy resolution to separate all characteristic spectra, completely eliminating line overlap, and affording high line to background ratios. Also, line position and shape would be independent of count rate effects. Since these attributes are not fulfilled completely, the X-ray analyst must judge beforehand the limitations and best procedures for quantitative analysis using energy spectrometry. On the other hand, rapid qualitative identification of major, and in most cases minor, chemical constituents is accomplished easily with present energy spectrometer resolution, and computer reduction can overcome many of the overlap problems.

Detection Efficiency

Figure 6 illustrates detection efficiency of a silicon detector for different depletion depths and beryllium window thicknesses. A windowless detector should have essentially 100 percent efficiency for all energies below 10 keV, to a lower limit set by electronic noise and detector dead layer (0.1 to 1.0 μm). The decreasing efficiency for higher energies ($>$ 10 keV) is caused by loss of photoelectric absorption due to a finite depletion depth. Germanium, on the other hand, has a higher photoelectric absorption coefficient for a given energy than silicon, and therefore a better detection efficiency at high energy for equal depletion depths. Unfortunately germanium can suffer a change in efficiency at its K absorption energy (11.1 keV) due to the probability of creating an escape peak and increased

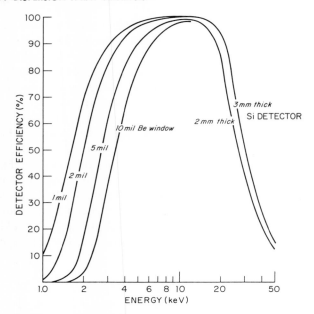

FIG. 6—*Detection efficiency of a silicon detector for different depletion depths and beryllium window thicknesses.*

absorption by the dead layer [*13*]. Thus, silicon is generally higher in efficiency than germanium in the 11.1 to 20.0 keV range and is preferred for X-ray spectrometry.

Resolution

A comparison of resolution between wavelength and energy spectrometers as a function of X-ray energies (0.1 to 36 keV) is given in Fig. 7. We have defined resolution as the full width at half height of a line divided by the energy of the line expressed as a percentage [*14*].

The measured resolutions of the wavelength spectrometers were obtained from an Applied Research Laboratories (ARL) microprobe and represent a practical compromise between resolution and count rate. This resolution is a variable which depends on types and perfection of the diffracting crystals and the geometry employed. The measured resolution of the ARL thin-window flow gas (90 percent argon, 10 percent methane) detector was obtained with a bias of 2050 V and a charge sensitive preamplifier. This curve is consistent with Eq 1 for $F = 1$ and thus follows Poisson statistics. A lower value of F might be obtained with optimized detector geometry and bias voltage. Gas detector characteristics are covered extensively by Sutfin and Ogilvie.[6] The resolution plot for the silicon energy spectrometer was calculated with Eq 1, $F = 0.13$, $\epsilon = 3.8$

[6] See p. 197.

eV/e^-h(77K), and a noise of 200 eV added in quadrature. This added electronic noise width accounts for the different slopes of the two energy spectrometers. While resolution of energy spectrometers improves with increasing energy, the converse is true for wavelength spectrometers. For energies above 18 keV, the silicon energy spectrometer has better resolution than the lithium fluoride spectrometer used in the comparison.

The measured lower spectrum in Fig. 8 illustrates the improved resolution obtained with a silicon detector at high X-ray energies (barium $K\alpha$ = 31.82, 32.19 keV; barium $K\beta$ = 36.34, 37.26 keV); the upper spectra were obtained from a wavelength spectrometer consisting of a flat lithium fluoride (LiF) diffracting crystal and a sodium iodide (NaI) scintillation counter. The 3rd order spectrum illustrates the inherent potential resolution of the wavelength spectrometer at higher energies at a cost of lower sensitivity. The quality of the solid-state detector can influence the resolution at high energy through distortion of line shape. Inadequate charge collection (trapping) results in asymmetrical line broadening (tailing)[15], which may be less apparent at lower energies. These effects are masked by electronic noise and less utilization of detector volume at low energies.

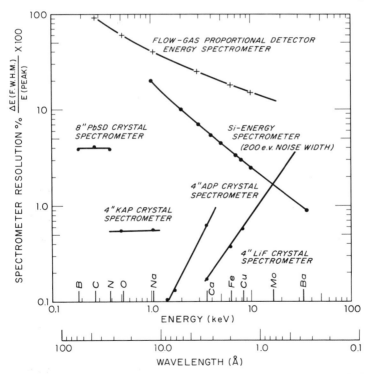

FIG. 7—*Comparison of resolution between wavelength and energy spectrometers as a function of X-ray energies.*

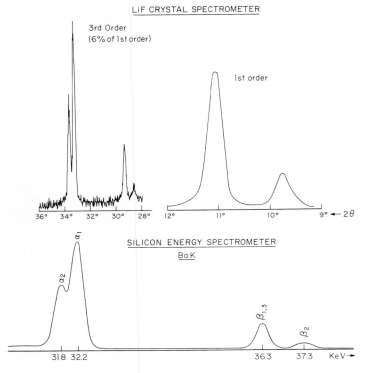

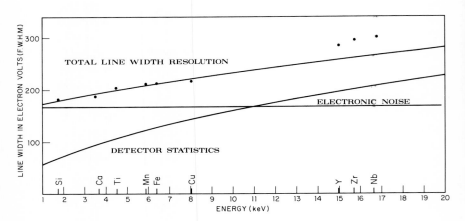

FIG. 8—*Comparison of wavelength and energy spectrometer resolution for barium K-radiation.*

FIG. 9—*Influence of electronic noise and detector statistics on resolution of energy spectrometer.*

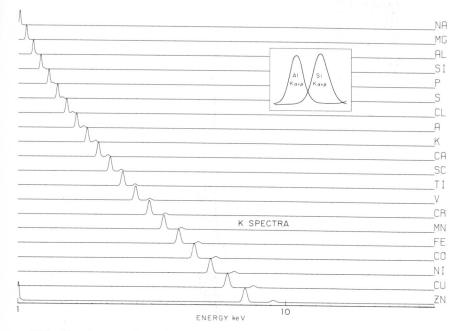

FIG. 10—*A comparison of measured aluminum and silicon* (insert) *and calculated zero electronic noise spectra. The measured spectra was obtained using a spectrometer with 166 eV of electronic noise.*

Electronic noise and detector statistics influence resolution in the 1 to 20 keV range as shown in Fig. 9. Since these combine as variances, and electronic noise is independent of energy, electronic noise dominates at the lowest X-ray energies. This can be seen in the experimental and theoretical curves of Fig. 9. The points represent measured $K\alpha$-line widths, and the electronic noise contribution was measured with a pulser (166 eV). In this, and in Figs. 10 and 11,[7] detector statistics were computed with $F = 0.13$ and $\epsilon = 3.8$ eV$/e^{-}$h in Eq 1. Good agreement in Fig. 9 between points and the predicted total width is maintained from 1 to 10 keV, and deviations for yttrium, zirconium, and niobium are due to significant separation of $K\alpha_1$- and $K\alpha_2$-lines.

The dominating influence of electronic noise is best illustrated by comparing measured aluminum and silicon spectra (insert, Fig. 10) with the highly hypothetical zero electronic noise spectra of Fig. 10.

Elemental identification by visual inspection of the oscilloscope is aided by the calculated spectra shown in Fig. 11 and listed in Table 1. The shape of a line, its position on the screen, and the relative intensity of the lines

[7] A full scale print of Fig. 11 (STP 485 X) is available from ASTM Headquarters at a cost of $2.00. Order No. 04-485025-39.

TABLE 1—Prominent X-ray line energies (keV).

Z		$K\alpha$	$K\beta$
11	NA	1.041	1.071
12	MG	1.254	1.302
13	AL	1.487	1.557
14	SI	1.740	1.836
15	P	2.013	2.139
16	S	2.307	2.464
17	CL	2.622	2.816
18	A	2.957	3.190
19	K	3.313	3.590
20	CA	3.690	4.013
21	SC	4.089	4.460
22	TI	4.509	4.932
23	V	4.950	5.427
24	CR	5.412	5.947
25	MN	5.895	6.490
26	FE	6.400	7.058
27	CO	6.925	7.649
28	NI	7.472	8.265
29	CU	8.041	8.905

Z		$L\alpha,\beta$	$K\alpha$	$K\beta$
30	ZN	1.019	8.631	9.572
31	GA	1.107	9.243	10.262
32	GE	1.198	9.876	10.980
33	AS	1.294	10.532	11.723

Z		$M\alpha,\beta$	Ll	$L\alpha_1$	$L\beta_1$	$L\beta_2$	$L\beta_3$	$L\gamma_1$	$L\gamma_2$
61	PM	1.036	4.814	5.432	5.961	6.339		6.892	7.176
62	SM	1.090	4.994	5.636	6.206	6.586		7.180	7.471
63	EU	1.142	5.177	5.846	6.456	6.843		7.480	7.768
64	GD	1.197	5.362	6.057	6.713	7.103		7.786	8.087
65	TB	1.253	5.547	6.273	6.978	7.367		8.102	8.398
66	DY	1.309	5.743	6.495	7.248	7.636		8.419	8.714
67	HO	1.365	5.943	6.720	7.525	7.911		8.747	9.051
68	ER	1.424	6.152	6.949	7.811	8.189		9.089	9.385
69	TM	1.482	6.342	7.180	8.101	8.468		9.426	9.730
70	YB	1.543	6.546	7.416	8.402	8.759		9.780	10.090
71	LU	1.606	6.753	7.656	8.709	9.049		10.143	10.460
72	HF	1.671	6.960	7.899	9.023	9.347		10.516	10.833
73	TA	1.738	7.173	8.146	9.343	9.652		10.895	11.217
74	W	1.805	7.388	8.398	9.672	9.961		11.286	11.608
75	RE	1.874	7.604	8.652	10.010	10.275		11.685	12.010
76	OS	1.944	7.822	8.912	10.355	10.598		12.095	12.422
77	IR	2.016	8.046	9.175	10.708	10.920		12.513	12.842
78	PT	2.088	8.268	9.442	11.071	11.250	11.610	12.942	13.270
79	AU	2.163	8.494	9.713	11.442	11.585	11.995	13.382	13.709
80	HG	2.238	8.721	9.989	11.823	11.924	12.390	13.830	14.162
81	TL	2.315	8.953	10.269	12.213	12.271	12.793	14.292	14.625
82	PB	2.393	9.184	10.551	12.614	12.623	12.793	14.764	15.101
83	BI	2.473	9.420	10.839	13.023	12.980	13.210	15.248	15.582
84	PO	2.553	9.664	11.131	13.447	13.340	13.638	15.744	16.070
85	AT	2.635	9.904	11.427	13.876	13.710	14.067	16.251	16.580
86	RN	2.714	10.143	11.727	14.316	14.080	14.512	16.770	17.120

34	SE	1.392	11.209	12.493	87	FR	2.795	10.383	12.031	14.770	14.450	14.976	17.303	17.659
35	BR	1.496	11.909	13.288	88	RA	2.875	10.622	12.340	15.236	14.841	15.445	17.849	18.179
36	KR	1.603	12.632	14.108	89	AC	2.972	10.870	12.652	15.713	15.233	15.931	18.405	18.730
37	RB	1.713	13.375	14.956	90	TH	3.069	11.119	12.969	16.202	15.624	16.426	18.982	19.305
38	SR	1.828	14.143	15.830	91	PA	3.158	11.366	13.291	16.702	16.024	16.930	19.568	19.872
39	Y	1.947	14.933	16.732	92	U	3.251	11.618	13.615	17.220	16.428	17.455	20.167	20.485
40	ZR	2.070	15.747	17.661	93	NP	3.345	11.870	13.944	17.750	16.840	17.989	20.785	21.099
41	NB	2.196	16.584	18.614	94	PU	3.433	12.124	14.279	18.294	17.255	18.540	21.417	21.725
42	MO	2.327	17.444	19.599	95	AM	3.522	12.384	14.618	18.852	17.677	19.106	22.065	22.361
43	TC	2.462	18.328	20.609										
44	RU	2.600	19.236	21.646										
45	RH	2.743	20.169	22.711										
46	PD	2.889	21.125	23.805										
47	AG	3.040	22.105	24.927										
48	CD	3.195	23.110	26.078										
49	IN	3.354	24.140	27.257										

		Ll	$L\alpha_1$	$L\beta_1$	$L\beta_2$	$L\gamma_1$	$L\gamma_2$	$K\alpha_2$	$K\alpha_1$	$K\beta_{1,3}$	$K\beta_2$
50	SN	3.045	3.444	3.663	3.905	4.131	4.377	25.196		28.465	29.109
51	SB	3.189	3.605	3.844	4.101	4.348	4.600	26.276		29.702	30.389
52	TE	3.336	3.769	4.030	4.302	4.571	4.829	27.382		30.970	31.700
53	I	3.485	3.938	4.221	4.508	4.801	5.066	28.514		32.267	33.042
54	XE	3.640	4.110	4.420	4.722	5.041	5.286	29.672		33.593	34.415
55	CS	3.795	4.286	4.620	4.936	5.280	5.542	30.625	30.973	34.953	35.822
56	BA	3.854	4.466	4.828	5.156	5.531	5.797	31.817	32.194	36.341	37.257
57	LA	4.124	4.651	5.042	5.384	5.788	6.060	33.034	33.442	37.761	38.730
58	CE	4.287	4.840	5.262	5.613	6.052	6.325	34.279	34.720	39.214	40.233
59	PR	4.453	5.034	5.489	5.850	6.322	6.598	35.550	36.026	40.701	41.773
60	ND	4.633	5.230	5.722	6.089	6.602	6.883	36.847	37.361	42.219	43.350

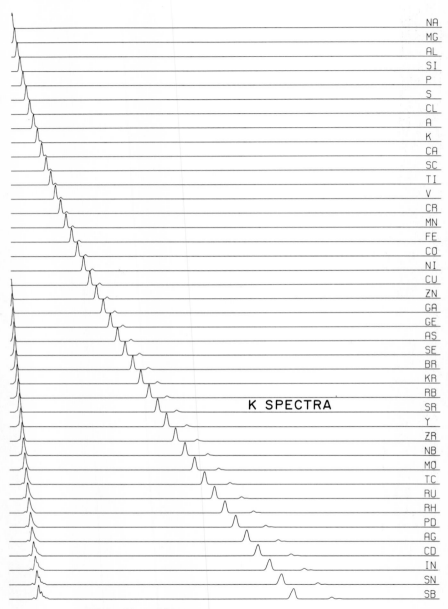

K SPECTRA

Weighted intensities are:

$K\alpha_1$ (1.0), $K\alpha_2$ (0.5), $K\beta_1$ (0.15), $K\beta_2$ (0.05), $K\beta_3$ (0.15)

Ll (0.05), $L\alpha_1$ (1.0), $L\alpha_2$ (0.1), $L\beta_1$ (0.5), $L\beta_2$ (0.2), $L\beta_3$ (0.06)

$L\gamma_1$ (0.1)

$M\alpha_{1,2}$ (1.0), $M\beta$ (0.80)

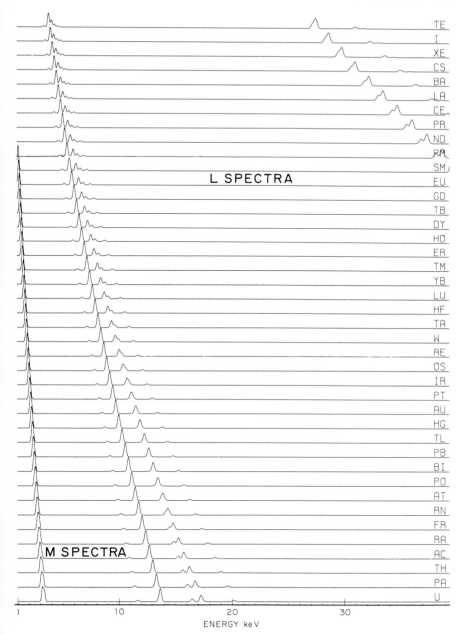

Calculated assuming 150 eV (FWHM) electronic noise and $\epsilon = 3.8$ eV/e⁻h, F = 0.13.

FIG. 11—*Theoretical characteristic X-ray spectra (atomic numbers 11 to 92).*

in a series, in most cases, can give enough visual information to the X-ray analyst for rapid identification. These theoretical characteristic X-ray spectra are for elements of atomic number 11 to 92 (1 to 39 keV) and have been given weighted intensities that are indicated in parentheses:

$$K\alpha_1 \ (1.0), \ K\alpha_2 \ (0.5), \ K\beta_1 \ (0.15), \ K\beta_2 \ (0.05), \ K\beta_3 \ (0.15)$$

$$L_l \ (0.05), \ L\alpha_1 \ (1.0), \ L\alpha_2 \ (0.1), \ L\beta_1 \ (0.5), \ L\beta_2 \ (0.2), \ L\beta_3 \ (0.06),$$

$$L\gamma_1 \ (0.1), \ M\alpha_{1,2} \ (1.0), \ M\beta \ (0.80)$$

The spectra were generated by the addition of individual Gaussian distributions for each line. The width (FWHM) of each distribution was calculated by Eq 2 assuming an electronic noise of 150 eV. Each K, L, and M series was normalized to its most prominent line (weight of one = full scale). Therefore, the spectra do not preserve relative intensities between the different X-ray series or include detection and generation efficiency, or matrix effects. A spectrometer with the resolution illustrated in Fig. 11 will have a 200 eV width (FWHM) for iron $K\alpha$ (6.4 keV).

For $K\alpha$-lines of equal intensity, peaks are separated clearly from sodium to praseodymium at this resolution. The adjacent element line overlap decreases with increasing atomic number, complete separation being obtained for elements higher than manganese. Spectral interference on $K\alpha$-lines by one atomic number lower $K\beta$-line is present from sodium to zinc, two atomic numbers from nickel to rhodium and three after rhodium. K-line interference from L- and M-spectra from the higher atomic number elements is more prominent than $K\beta$ and due to their intensity presents a more severe overlap problem. In analytical problems where line interference is encountered, the use of spectrum stripping is necessary.

Count Rate and Line/Background

The intensities and line/background ratios that can be obtained are greatly influenced by the type of excitation employed and the resolution of the energy spectrometer. This is demonstrated in Figs. 12 and 13, which were made from electron and X-ray excitation, respectively. The lower line/background ratios in Fig. 12 are due to the continuous radiation (bremsstrahlung) that results from electron excitation. While this continuum is also incident upon the specimen in fluorescence excitation (Fig. 13), only a small fraction is backscattered to the detector (except at high energies with low Z materials).

The electron excitation data were obtained with an ARL microprobe and a 210-eV (FWHM iron $K\alpha$) ORTEC energy spectrometer. The detector is mounted on an existing X-ray port 27 cm from the specimen at an X-ray emergence angle of 52.5 deg. The active diameter and depth are 4 and 3 mm, respectively. The detector is collimated to 2 mm diameter and separated from the microprobe vacuum system by a 2 mil beryllium

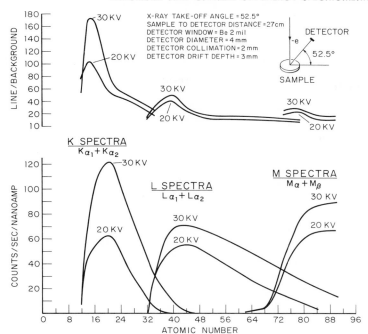

FIG. 12—*Intensities and line/background for electron excitation.*

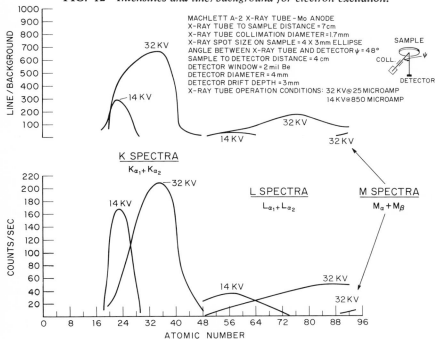

FIG. 13—*Intensities and line/background for X-ray excitation.*

window. Count rates correspond to total integrated intensities and line/ background is measured from peak channel intensities. The data for the K- and L-spectra are the sum of α_1- and α_2-lines, whereas the M-spectra are the sum of $\alpha_{1,2}$ and β. The X-ray fluorescence data in Fig. 13 were obtained from close coupling a Machlett A-2 molybdenum target X-ray tube, specimen holder, and 220 eV (FWHM iron $K\alpha$) ORTEC energy spectrometer. With this arrangement, high counting rates were obtained with microamperes rather than milliamperes X-ray tube anode current. However, poor sensitivity for elements below $Z = 18$ resulted from absorption in a 4-cm air path between the specimen and detector. With long integration periods this air path adds an argon $K\alpha$-peak to the observed spectrum. A specially designed vacuum cryostat could reduce the air path to a 5 mm length, thereby allowing 60 percent transmission of silicon $K\alpha$-radiation without enclosing the specimen in a vacuum. Because helium leakage into cryogenically pumped detector systems will degrade the vacuum, a helium X-ray path is only feasible for perfectly sealed systems, or those employing an ion pump with sufficient helium pumping speed.

Diffraction effects

The close coupling which is possible with the ES allows an enormous variation (10^4) in anode current of a conventional X-ray tube.[8] This permits reduction of the tube voltage and increase of total power in order to discriminate against higher energies while enhancing lower energies. However, lower excitation voltage emphasizes in our case ($\theta = 66$ deg) diffraction of incident continuous radiation by crystalline specimens to the point where interfering lines appear. This situation forms an interesting contrast to experiments of Giessen and Gordon [16] who employed $\theta = 11$ deg. In their situation, fluorescent radiation suffers greater absorption than the higher energy coherently diffracted beams. For comparable d-spacings, our $\theta = 66$ deg has shifted diffraction lines to lower energies and lower intensities relative to fluorescent lines. A large fraction of the incident continuum that exceeds the absorption edge energy contributes to a fluorescent line, while only a small fraction is utilized by the diffraction line. Figure 14 illustrates diffraction and characteristic X-rays for two orientations of a germanium single crystal. In this case, the diffraction lines have been exaggerated by using a low excitation potential to suppress germanium $K\alpha$-lines.

In contrast to the interference posed by diffraction lines in X-ray fluorescence, the opposite situation exists in conventional X-ray diffraction. Solid-state X-ray detectors exhibit sufficient energy resolution to completely discriminate against specimen fluorescence and eliminate the need for

[8] Dark currents in sealed X-ray tubes may prevent their use in the low anode current ranges.

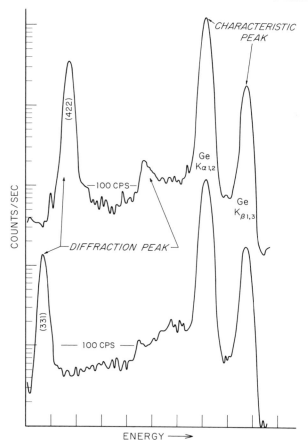

FIG. 14—*Simultaneous detection of diffraction and characteristic peaks using X-ray excitation.*

the $K\beta$ filter [17]. The detector provides line to background ratios similar to those obtained using a focusing monochromator and peak intensities greater than can be obtained from a gas filled proportional counter with a $K\beta$ filter. Pictured in Fig. 15 is a solid-state detector which has a rectangular active area (19 by 1.5 mm) to conform to conventional collimation of the General Electric XRD-5 diffractometer.

The solid-state detector in conjunction with the single channel analyzer form the discrimination system necessary for diffraction recording; the addition of multichannel recording makes possible simultaneous fluorescence analysis.

Figure 16 shows a series of diffraction scans over the 3.14 Å reflection of silicon (28.4 deg 2θ for copper $K\alpha$; 25.5 deg 2θ for copper $K\beta$). The peak between the copper $K\alpha$- and copper $K\beta$-peaks is a tungsten $L\alpha$-peak

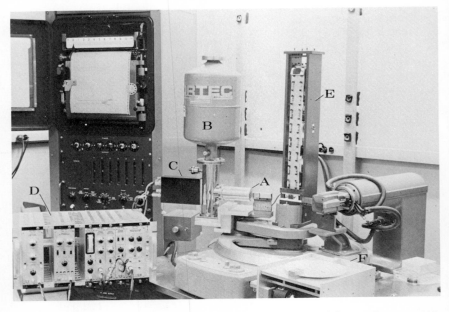

FIG. 15—*Conventional X-ray diffraction using a silicon solid-state detector.* (A) *Detector window and active area.* (B) *Liquid nitrogen dewar.* (C) *Preamplifier.* (D) *Power supply, baseline restorer, amplifier, single-channel analyzers, and ratemeter. Also shown are automatic specimen changer* (E) *and programming device* (F).

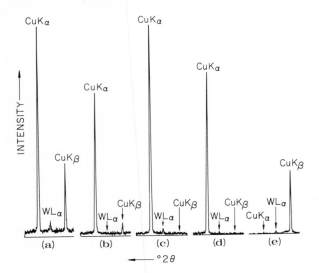

FIG. 16—*Diffraction scans of the (111) peak of silicon (3.14Å).* (a) *No pulse height discrimination or filter.* (b) *10 μ nickel Kβ filter, no pulse height discrimination.* (c) *Pulse height analyzer set to eliminate copper Kβ, no nickel filter.* (d) *Pulse height analyzer set to eliminate copper Kβ and WLα, no nickel filter.* (e) *Pulse height analyzer set to receive copper Kβ only.*

resulting from contamination of the anode of the X-ray tube. In Fig. 16*a*
no filter or pulse height discrimination is used.

A comparison between Fig. 16*a* and *b* illustrates the effect of a nickel
filter of the type commonly used for reducing the intensity of the copper
Kβ radiation. In Fig. 16*c* energy discrimination is used without the *Kβ*
filter. At this setting the tungsten *Lα*-line is reduced but still present; it
can be eliminated completely by energy discrimination (Fig. 16*d*), but this
results in some reduction in copper *Kα* intensity. It is also possible to obtain a pure copper *Kβ*-line (Fig. 16*e*), which may be useful in some diffraction problems where the splitting of the $K\alpha_1\alpha_2$ doublet makes the use of
Kα-radiation undesirable. It is also apparent from Fig. 16 that the use of a
narrow energy window results in a reduced background, and hence higher
line-to-background ratios.

Figure 17 illustrates the improvement in line/background ratio obtainable from an iron-rich specimen (ferric oxide (Fe_2O_3)). In Fig. 17*a* no
energy discrimination was used. In Fig. 17*b* the resolution of the detector
was reduced electronically to that obtainable with a gas proportional detector, and a nickel *Kβ* filter was employed. In Fig. 17*c* energy discrimination was used to eliminate all radiation except copper *Kα*. An added benefit
is the inherent long term pulse height stability of solid-state detectors.
Thermal isolation of the detector and first amplifier greatly reduces the influence of external temperature variations, which degrade gas detectors.

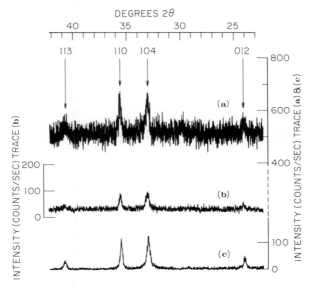

FIG. 17—*Diffraction scans over a natural hematite (α-Fe$_2$O$_3$) specimen. (a) Without pulse height discrimination or filter. (b) With nickel filter and pulse height discrimination with energy resolution similar to that obtainable with a gas proportional
detector. (c) With pulse height discrimination (resolution 390 eV).*

Sample Positioning Effects

An important characteristic of the energy spectrometer that contrasts with the wavelength spectrometer is its lack of sensitivity to X-ray source position. In situations where the specimen position may vary, more reliable results can be obtained from energy spectrometry. In the electron microprobe, for example, differences in specimen and standard elevations cause significant intensity errors from wavelength spectrometers (Fig. 18). In this case, the lack of sensitivity of the ES to specimen elevation is the result of a large specimen to detector distance.

This insensitivity can be also obtained with close coupling in X-ray fluorescence. By taking advantage of anode absorption characteristics at low takeoff angle, the auto-focusing arrangement of detector (*A*), specimen (*B*), source (*C*) can be devised, shown in Fig. 19. The instrument in Fig. 19 is designed for qualitative and semiquantitative analysis of deep-sea sediment cores. The cores vary from 10 to 150 cm in length and are sectioned along their major axis. The dynamically pumped demountable X-ray tube is an Ehrenburg-Spear [*18*] design and is operated at 50 kV at 1 μA anode current. The X-ray illumination on the specimen is an ellipse of varying intensity along the major axis. The variation in intensity coupled with detector collimation cancels the loss of intensity when the specimen is moved away from the detector. A change in distance of about ±2.5 mm gives no appreciable change in counting rate for detector to specimen distance of 2.5 cm.

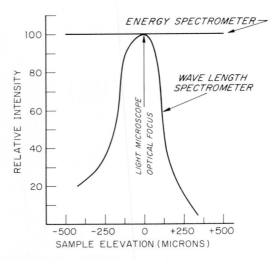

FIG. 18—*Comparison of relative intensity between an energy spectrometer and wavelength spectrometer for variation in specimen elevation in microprobe analysis.*

FIG. 19—*X-ray fluorescence instrumentation for the analysis of long cores of deep sea sediments. (A) Solid-state (Si) detector and cryostat. (B) Specimen (long core) (C) X-ray tube.*

Data Handling and Reduction

Energy spectrometer data stored in a multichannel analyzer are ideally suited for digital computations, but their rate of generation and storage requirements necessitate special handling. With the short integration periods possible, a plethora of spectra is produced by an ES in a short time. Good peak delineation is obtained if five channels describe the FWHM, and this results in 300 channels to cover a 15 keV range for FWHM of 250 eV. Thirty spectra thus imply 9000 (6 digit) numbers, more than enough to inundate the operator in yards of punched or printed paper tape. Possible solutions to this problem are given below in order of increasing cost.

The oscilloscope display of a multichannel analyzer provides an immediate visual "fingerprint," and accurate quantitative comparisons can be made between two spectra in separate memory sections. Rapid individual peak identification in a complex spectrum is best served by a single control used to position an intensified indicator through the display coupled to digital readout of the indicated position (channel number).

An X-Y plot and printed output for brass are given in Fig. 20. For the isolated copper $K\alpha$-peak, the printed output gives all the necessary information for peak location, subtraction of background, and numerical integration. However, the overlapping zinc $K\alpha$ and copper $K\beta$ distributions are better observed by an X-Y plotter, and peak heights and locations are easily measured. Myklebust and Heinrich [19] have shown that an analog curve resolver may be used for separation and integration of complex peaks of an X-Y plot made at two attenuations (for both large and small peaks). Overlapping peak distributions should be considered the rule, not the exception, especially in microprobe or SEM work below 20 keV.

Integrators can be obtained for summing successive channels of a multichannel analyzer, and there is one currently available which will integrate over several selected energy ranges in a single pass.

Complete reduction of large amounts of spectral data into energies and amplitudes can be accomplished only with a digital computer. Probably the least expensive way to communicate with a computing facility is by magnetic tape output, but processing is then delayed by hours or even days. Replacement of the multichannel analyzer by a computer (Fig. 21e) can provide a large degree of flexibility in oscilloscope display and data reduction, enabling critical evaluation of residuals produced by peak subtraction or checking the results of peak searching techniques. Once the

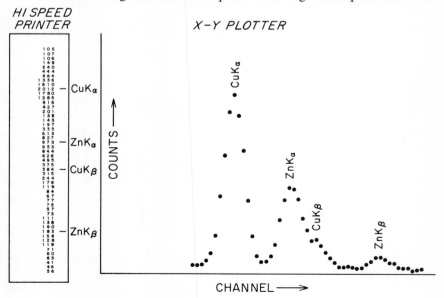

FIG. 20—*Types of output from multichannel analyzers.*

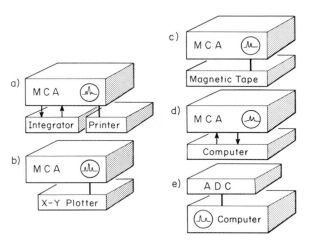

FIG. 21—*Instrumentation for data accumulation, handling and reduction for energy spectrometers.*

digital data are stored in the computer, reduction involves calculation of peak locations, amplitudes, and subtraction of background counts. Savitzgy and Golay [20] have tabulated coefficients for smoothing spectra or their derivatives; three examples are shown in Fig. 22 for a samarium L-spectrum. Whatever function is used for peak searching, a major problem is setting significance tests that will discriminate against background fluctuations without eliminating small peaks.

We have found that each X-ray energy produces a Gaussian-shaped distribution that can be fitted by ordinary least squares procedures.

$$C_i = A + Bx_i + \sum_j D_j \exp \left[-(x_j - x_i)^2 E_j\right] \ldots\ldots\ldots\ldots(3)$$

expresses the counts, C_i, in the ith channel, x_i, as a sum of linear[9] background (A,B) parameters and j Gaussian peaks $(D_j = $ amplitude, $x_j = $ peak position, $E_j = $ width parameter). The set of linear equations

$$C_i^o - C_i^c = \Delta A + x_i \Delta B + \sum_j [\Delta D_j - 2(x_j - x_i)D_j E_j \Delta x_j$$
$$- (x_j - x_i)^2 D_j \Delta E_j] \exp \left[-(x_j - x_i)^2 E_j\right]$$

are solved for the parameter shifts.

$(\Delta A, \Delta B, \Delta D_j, \Delta x_j, \Delta E_j)$ which minimize the quantity

$$\sum_i (C_i^o - C_i^c)^2 / C_i^o$$

where C_i^o are the observed counts and C_i^c is calculated from Eq 3. Since the weighting scheme, $1/\sqrt{C_i^o}$, gives a rather embarrassing value for a channel

[9] If significant curvature in background is present, then a higher degree more general function can be fitted experimentally. Parameters of this function should be restricted to give an expected form thereby preventing arbitrary solutions.

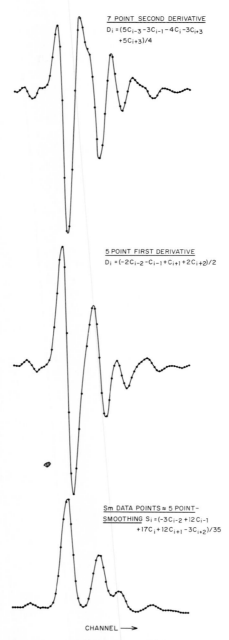

7 POINT SECOND DERIVATIVE
$$D_i = (5C_{i-3} - 3C_{i-1} - 4C_i - 3C_{i+3} + 5C_{i+3})/4$$

5 POINT FIRST DERIVATIVE
$$D_i = (-2C_{i-2} - C_{i-1} + C_{i+1} + 2C_{i+2})/2$$

Sm DATA POINTS ≈ 5 POINT–
SMOOTHING $S_i = (-3C_{i-2} + 12C_{i-1} + 17C_i + 12C_{i+1} - 3C_{i+2})/35$

CHANNEL $\longrightarrow$

FIG. 22—*Tabulated coefficients for smoothing spectra or their derivatives.*

containing zero counts, a weight of 1 is used in this instance. Several cycles of least squares are required for a given set of peaks because Eq 3 is not linear. If the width parameters, E_j, are refined either individually or collectively (by keeping E_j fixed and factoring out a single variable multiplier e^E), the initial guess must be fairly close to the correct answer in order to avoid oscillating or even diverging parameter shifts. Once peak width as a function of energy and counting rate has been established for a given energy spectrometer system, restriction of this parameter leads to very rapid convergence and a method for checking inconsistencies in the results.

The tellurium spectrum in Fig. 23 is shown before and after various numbers of peaks have been subtracted by the least squares fitting procedure. Six peaks are necessary in order to eliminate significant systematic deviations from background. Subtraction of all peaks of a *major* constituent might employ fixed line positions and relative intensities in order to reveal badly overlapped smaller peaks.

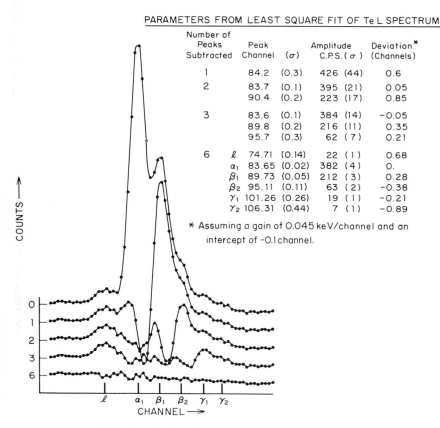

PARAMETERS FROM LEAST SQUARE FIT OF Te L SPECTRUM

Number of Peaks Subtracted		Peak Channel	(σ)	Amplitude C.P.S. (σ)	Deviation* (Channels)
1		84.2	(0.3)	426 (44)	0.6
2		83.7	(0.1)	395 (21)	0.05
		90.4	(0.2)	223 (17)	0.85
3		83.6	(0.1)	384 (14)	−0.05
		89.8	(0.2)	216 (11)	0.35
		95.7	(0.3)	62 (7)	0.21
6	ℓ	74.71	(0.14)	22 (1)	0.68
	α_1	83.65	(0.02)	382 (4)	0.
	β_1	89.73	(0.05)	212 (3)	0.28
	β_2	95.11	(0.11)	63 (2)	−0.38
	γ_1	101.26	(0.26)	19 (1)	−0.21
	γ_2	106.31	(0.44)	7 (1)	−0.89

* Assuming a gain of 0.045 keV/channel and an intercept of −0.1 channel.

FIG. 23—*Peak subtraction by least square fitting.*

Experience which we have had in least squares fitting of Gaussian peaks to overlapping spectra suggests that automatic data reduction is feasible. The visual criteria that we have employed for obtaining peak position and amplitude estimates can be expressed analytically in terms of peak shape and significant size. Just as the linear amplifier extracts signal from noise, mathematical filtering of a spectrum can be accomplished through tests that distinguish the longer period undulations in background from peaks of appropriate FWHM. Special problems in background such as beryllium window and specimen absorption of low energy bremsstrahlung in the microprobe can be handled specifically.

Analytical problems consisting of many specimens of similar composition, where a sufficient number of standards can be made to bracket the compositional variations of each element, may be handled by "spectra matching." Massive files of standard spectra can be stored on magnetic disk or tape and computer recalled by means of file identification derived from the characteristic of the unknown spectrum. A complete spectrum comparison can be rapidly obtained by providing information on the degree of fit for quantitative analysis. Interpolation between sets of standard spectra to establish working curves can extend the analytical range.

Acknowledgments

The authors thank G. Arrhenius, J. Frazer, J. I. Drever, and B. Swope for contribution to the development of the project. Handling of spectra and digital reduction were conducted at Gulf General Atomic Incorporated with vital assistance from E. Anderson and B. Cross. We thank the Planetology Branch of NASA—Ames Research Center, Moffett Field, California, for use of their multichannel analyzer. The first author acknowledges the support from Oceans Exploration Division of Kennecott Exploration Inc., and from the Air Force Office of Scientific Research under grant number AF-AFOSR-631-67-A.

References

[1] Buck, T. M. in *Semiconductor Nuclear-Particle Detectors and Circuits,* Brown, W. L., ed., National Academy of Science, National Research Council Publication 1593, Washington, D. C., 1969, p. 144.

[2] Dearnaley, G. in *Semiconductor Nuclear-Particle Detectors and Circuits,* Brown, W. L., ed., National Academy of Science, National Research Council Publication 1593, Washington, D. C., 1969, p. 5.

[3] Fano, U., *Physical Review,* Vol. 72, No. 1, July 1947, p. 26.

[4] Bilger, H. R. in *Semiconductor Nuclear-Particle Detector and Circuits,* Brown, W. L., ed., National Academy of Science, National Research Council Publication 1593, Washington, D. C., 1969, p. 50.

[5] Walter, F. J. in *Semiconductor Nuclear-Particle Detector and Circuits,* Brown, W. L., ed., National Academy of Science, National Research Council Publication 1593, Washington, D. C., 1969, p. 63.

[6] McKay, K. G., *Physical Review,* Vol. 76, No. 10, 1949, p. 1537.

[7] Pell, E. M., *Journal of Applied Physics,* Vol. 31, No. 2, Feb. 1960, p. 291.

[8] Elad, E., *Nuclear Instruments and Methods,* Vol. 37, 1965, p. 327.
[9] Radeka, V. in *Semiconductor Nuclear-Particle Detector and Circuits,* Brown, W. L., ed., National Academy of Science, National Research Council Publication 1593, Washington, D. C., 1969, p. 393.
[10] Goulding, F. S. in *Semiconductor Nuclear-Particle Detector and Circuits,* Brown, W. L., ed., National Academy of Science, National Research Council Publication 1593, Washington, D. C., 1969, 381.
[11] Wilkinson, D. H., *Proceedings of the Cambridge Philosophical Society,* Vol. 46, 1950, p. 508.
[12] Sutfin, L. V., Ogilvie, R. E., and Harris, R. S., presented at *Fourth National Conference on Electron Microprobe Analysis,* July 1969, Pasadena, Calif.
[13] Walter, F. J., *IEEE Transactions,* Nuclear Science 12th Scintillation and Semiconductor Counter Symposium, 1970 (in press).
[14] Siegbahn, K., *Beta and Gamma-Ray Spectroscopy,* North-Holland Publishing Co., Amsterdam, 1955, p. 56.
[15] Mayer, J. W. in *Semiconductor Nuclear-Particle Detectors and Circuits,* Brown, W. L., ed., National Academy of Science, National Research Council Publication 1593, Washington, D. C., 1949, p. 88.
[16] Giessen, B. C. and Gordon, G. E., *Science,* Vol. 159, No. 3818, March 1968, p. 973.
[17] Drever, J. I. and Fitzgerald, R. W., *Materials Research Bulletin,* Vol. 5, 1970, pp. 101-108.
[18] Ehrenberg, W. and Spear, W. E., *Proceedings of the Physical Society,* Vol. B64, 1951, pp. 67-75.
[19] Myklebust, R. L. and Heinrich, K. F. J., presented at *Fourth National Conference on Electron Microprobe Analysis,* July 1969, Pasadena, Calif.
[20] Savitzky, A. and Golay, J. E., *Analytical Chemistry,* Vol. 36, No. 8, 1964, p. 1627.

D. W. Aitken[1] and E. Woo[2]

The Future of Silicon X-ray Detectors*

REFERENCE: Aitken, D. W. and Woo, E., "The Future of Silicon X-ray Detectors," *Energy Dispersion X-ray Analysis: X-ray and Electron Probe Analysis*, *ASTM STP 485*, American Society for Testing and Materials, 1971, pp. 36–56.

ABSTRACT: The future potential for silicon X-ray detectors is appraised in view of recent accomplishments, present trends, and present physical knowledge. The discussion includes achievements in resolution with both small and large area detectors, and at both low and high count rates. Differences are emphasized between theoretical capability and experimental practicality, with the dual aim of helping the experimenter to evaluate the potential of silicon X-ray spectrometers in his own work, and of guiding him toward experimental design concepts that might enable him to make the maximum use of the available electronic and detector performance capabilities. The present advantages of silicon over germanium for X-ray energies below about 40 keV are discussed.

KEY WORDS: solid state counters, silicon, germanium, spectrometers, X-ray spectra, X-ray spectrometers, X-ray fluorescence, energy absorption, resolution, dispersion, radiation measuring instruments

The most straightforward way for inferring what might be in store for silicon X-ray detectors is to evaluate recent developments and applications which, in large measure, will help to define future trends. For that reason we have chosen first to describe the present status of energy-dispersive X-ray spectrometry, and then to take a *practical* look at where we should try to go with this technology in the future.

The reader can find useful additional reference material and bibliographic assistance in a somewhat more balanced presentation in the recent reviews by Walter [1][3,4] and Frankel and Aitken [2]. More specific and comprehensive comparisons of different types of energy-dispersive radiation detectors can be found in the 1967 review by Burkhalter and Camp-

---

* Research supported in part by the Air Force Office of Scientific Research, Office of Aerospace Research, U. S. Air Force under AFOSR Contract F 44(620) 69-C-0042.

[1] Chairman, Department of Environmental Studies, San Jose State College, San Jose, Calif.

[2] Technical director, the Kevex Corporation, Burlingame, Calif.

[3] The italic number in brackets refer to the list of references appended to this paper.

[4] See p. 82.

bell [3], the 1968 review by Aitken [4], and the 1969 review by Campbell, Burkhalter, and Marr [5].

Silicon X-ray Spectrometers: Present and Future

Present State of the Art

Practical View of High Resolution—The surge of interest in energy-dispersive X-ray emission spectroscopy that has been in evidence during the past four years has been sparked by the extraordinary improvement in resolution that has been realized with semiconductor radiation detectors during that period. This is amply demonstrated in Fig. 1, which illustrates the comparative response of three different popular energy-dispersive radiation detectors to the manganese X-ray spectrum emitted by the electroncapture isotope iron-55 (Fe^{55}).

Another Fe^{55} spectrum is illustrated in Fig. 2, demonstrating the exceptional performance that now can be realized with silicon X-ray spectrometers, not just as isolated examples in the laboratory, but consistently with commercially available apparatus.

We must caution, though, that features which are not readily apparent to the eye determine the true worth of such accomplishments. The full width at tenth maximum (FWTM), for example, better reveals the performance of the system than does the full width at half maximum (FWHM), since not only must the resolution be good, but the spectrum must be free of tailing and distortions. A good Gaussian reproduction of a

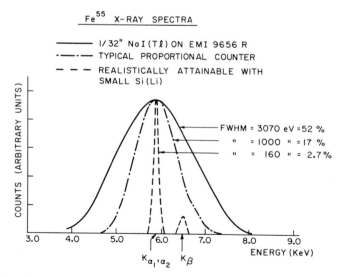

FIG. 1—*Comparison of the response of various energy-dispersive X-ray detectors to the manganese X-ray spectrum resulting from the decay of Fe^{55}.*

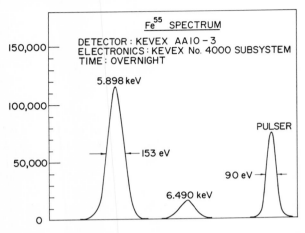

FIG. 2—*An actual Fe^{55} spectrum, showing the long-run performance that can now be attained with commercially available apparatus. While the full width at half maximum is shown to be 153 eV, the full width at tenth maximum was 283 eV, demonstrating an accurate Gaussian reproduction of the spectrum.*

monoenergetic X-ray source gives a value of about 1.85 for the FWTM/FWHM ratio. The spectrum illustrated in Fig. 2 demonstrated a FWTM of 283 eV, yielding the desirable ratio of 1.85 for the two width measurements.

As detectors and electronics continue to improve this would appear to suggest that one might have to go to absurd lengths to seek a more sensitive criterion of spectral quality. In the copper X-ray spectrum shown in Fig. 3, the full width at hundredth maximum is seen to be still above the region of substantial broadening due to tailing.

Actually, system resolutions of the quality demonstrated in Figs. 2 and 3 simply could not be achieved if there were any significant spectral distortions present, other than those fairly small effects that are related fundamentally to the physical processes within the detector material [6-8]. Therefore, the conclusion can be drawn that when very good resolution is demonstrated by a system, the FWHM is probably a fully adequate practical measure of the overall spectral quality.

The peak-to-valley ratio between adjacent peaks is another important measure of system resolution. In practice, though, this ceases to be a useful measure when peaks are resolved fully, as are the two X-ray spectra shown in Fig. 2. In fact, the peak-to-valley ratio ceases to be an adequate indication of spectral quality when the peaks are separated by more than about 3.5 times the FWHM. Under this condition the valley is determined by system background, rather than system resolution.

For example, while a peak-to-valley ratio for the spectrum shown in Fig. 2 was measured to be about 300:1, with the source positioned immediately adjacent to the beryllium window of the detector cryostat, it degraded to about 200:1 with the source removed by a few inches. This demonstrates that the peak-to-valley ratio is being determined not by system electronic performance, but rather by background conditions, such as air and window-scattering and edge effects. But this also demonstrates that, for comparison purposes, a standardized geometry should be adopted to yield at least uniform background conditions. For our purposes this meant always placing the radioactive source on or against the beryllium window.

Finally, one additional criterion related to system resolution that should be mentioned here is the effect of long counting times. For example, while Fig. 4 represents a short counting time, the spectra shown in Figs. 2 and 3 both represent what can be fairly called long counting times. Minor electronic drifts over the longer periods have degraded the resolution by about 5 eV. This was verified by measuring the corresponding short-run FWHM for the Fe[55] K-alpha line, which was less than 150 eV.

Count-Rate Capability—The improvement in the performance of the electronics at high count rates over the past four years probably has been as dramatic as the improvement in resolution, and certainly has been of comparable value in practical applications.

Table 1 illustrates the performance of the system utilized to produce the spectra shown in Figs. 2 and 3 as a function of the Gaussian pulse-shaping time constant and of total recorded count rate. Although some degradation of the resolution with count rate appears to be unavoidable, a 15 percent penalty in resolution with a 10-mm^2 detector, or a seven percent penalty with a 200-mm^2 detector, to achieve a count-rate capability of 20,000 counts per second (cps), is a fairly small price to pay for the corresponding great reduction in experimental time.

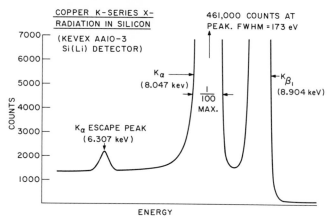

FIG. 3—*The fluorescent K-series X-radiation from copper, expanded to show the spectral quality at one percent of the peak height, as well as to show the presence of the escape peak in silicon.*

TABLE 1—*Resolution versus count rate and area[a].*

Counts per Second	System Time Constant, μs (Gaussian)	Total System Resolution, eV[b]		
		5.898 keV in 10 mm² Si(Li)	6.403 keV in 80 mm² Si(Li)	6.403 keV in 200 mm² Si(Li)
500	8	153	243	301
1 000	8	154	243	301
5 000	8	162	249	306
10 000	8	171	255	310
10 000	6	167	252	315
20 000	6	180	261	323
20 000	4	178	. . .	. . .
50 000	2	207	. . .	438
50 000	1	237	. . .	. . .
100 000	2	230	. . .	449
100 000	1	240	. . .	. . .

[a] Measurements with Kevex Grade AA detectors, 2002 preamplifier and 4001P amplifier, with baseline restoration and pileup rejection.

[b] Table entries omitted when recent measurements not available.

Detector Area—It is apparent that with such high resolution now possible with silicon X-ray spectrometers, greater latitude is provided the experimenter in choosing a system that meets specifications other than just resolution. For example, although system resolution is degraded somewhat when larger area detectors are used, improvements in the performance of such systems have at least paralleled the other improvements cited above.

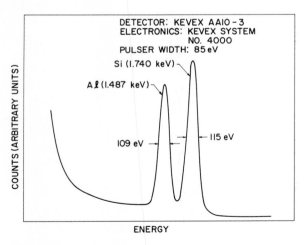

FIG. 4—*The fluorescent X-ray spectra from aluminum and silicon, adjacent elements on the periodic scale with, respectively, Z = 13 and Z = 14. Already at these energies the system is being limited in sensitivity more by window and air absorption than by electronic performance.*

This prompted us to indicate in Table 1 the performance of detectors varying by a factor of 20 in area. This particular subject is explored in detail in a paper now being prepared by Frankel and Aitken [9].

It is instructive to conclude this brief status description by noting that in the review published in 1968 by Aitken [4] the best (1967) resolution value available at the time of the writing of the manuscript was 330 eV, obtained at a low count rate with a 30-mm² detector. Table 1 in the present work demonstrates that that same value can now be achieved with a detector almost seven times as large operating at 20,000 cps.

Practical Look at the Future

Toward Better Resolution—It is useful to look at the practical value of seeking ever-better resolution with semiconductor X-ray spectrometers, and then to try and estimate just how far this might go. Toward this latter aim, though, about all we can do is to rely upon an extrapolation of the developments of the past few years, and upon the best physical evidence that we have today regarding the "ultimate" resolution capabilities of the detectors, in order to make intelligent guesses.

High resolution with very small detectors—Certainly there are no reasonable arguments against the practical advantage of seeking ever-better resolution, while there are powerful arguments for it. Higher resolution with a given size detector and at a given count rate always provides greater sensitivity and analytical convenience. The better the resolution, the shorter the time necessary to resolve a spectral feature of a given intensity, and the greater the sensitivity of the system to even weaker features, or to features lying at very low energies. We offer three illustrations to support this claim.

Figure 3 reveals the "escape" peak corresponding to the K-alpha X-ray peak from copper, lowered in energy by that of the escaping silicon K-alpha X-ray, 1.740 keV, and reduced in intensity by a factor of about 550. The general value of this illustration is to show the advantage of high resolution in observing weak spectral features in the presence of background and in the vicinity of much larger spectral features.

Figure 4 reveals the role of high resolution in enabling the separation of elemental constituents, lying adjacent on the periodic scale, down to ever lower energies, where the energy separation of adjacent X-ray spectra becomes ever less. The illustrated performance suggests that with this system the adjacent oxygen and fluorine spectra should be also resolvable with a peak-to-valley ratio of nearly 2:1; while the adjacent carbon and nitrogen X-ray spectra should both be observable and resolved with a peak-to-valley ratio of better than 1.3:1. The thick window and geometry of the system used prevented experimental confirmation of these conclusions. The actual detection of oxygen in fact, was, reported by Elad et al [10], using an earlier system which is not at all representative of the excellent performance that he is now obtaining.

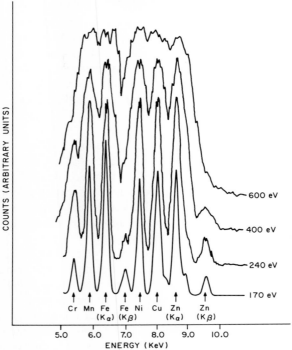

COMPARISON OF SYSTEM RESOLUTION
ON ADJACENT ELEMENTS
NOTES: BASE LINES SHIFTED FOR CLARITY.
NUMBERS SHOWN ARE FOR 5.9 KeV.

FIG. 5—*A graphic illustration of the value of high resolution in identifying the X-ray spectra from adjacent elements on the periodic scale, and in observing interferences between K-alpha and K-beta spectra of adjacent elements.*

Figure 5 serves the dual purpose of illustrating rather graphically the value of resolution in separating the X-ray spectra from adjacent elements, and also in providing for the separation of interfering spectra. For example, Fitzgerald and Gantzel note[5] that one can expect an interfering overlap of the K-alpha lines from elements spanning from sodium ($Z = 11$) to germanium ($Z = 32$) by the K-beta spectra from elements lying one atomic number lower. This is illustrated in Fig. 5, where the K-beta emission from nickel is seen to overlap the K-alpha spectrum from copper, and the K-beta X-rays from copper interfere with the K-alpha spectrum of zinc.

Figure 6 reveals the best reported resolution attained with a silicon detector on the Fe^{55} spectrum, as a function of year. It is evident that the rate of improvement is declining rapidly, and that some sort of asymptotic limit appears to be near. Two such possible limits have been drawn.

[5] See p. 3.

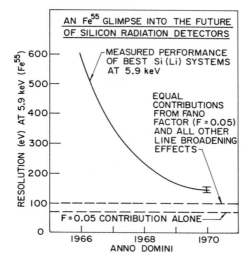

FIG. 6—*The improvement of silicon X-ray spectrometers with time, showing both the rapid recent improvement in resolution, and evidence for an approach to an ultimate asymptotic limit. The two possible dashed limits are discussed in the text.*

The lower limit drawn in Fig. 6 represents our best guess at present of the minimum theoretical resolution on the Fe^{55} K-alpha emission that can be calculated for silicon detectors. This was figured on the basis of Zulliger's estimate of 0.05 for the theoretical Fano factor for silicon [11] where the Fano factor represents a measure of the ultimate theoretical statistical capability of ionization detectors, and hence a kind of "figure of merit" for the performance of the detector element itself. This is clearly an unrealizable limit, since it does not take into account any other linebroadening influences upon the spectrum.

Assuming the accuracy of the Zulliger theory [11,12], a more realistic estimate might be to suggest that electronic improvements could lead to another factor of two improvement in electronic noise, while improvements in the quality of the silicon and further reductions in the size of the partially "dead" layer at the surface of the detector also might lead to a reduction in the "effective" Fano factor for silicon by close to a factor of two. Walter [1], Elad,[6] and Radeka [13] imply in their recent work that such an improvement in electronic performance might not be unreasonable to expect, while the recent remarkable work by the Berkeley group has led to an upper experimental limit for the Fano factor in germanium of 0.075 [14], or 32 percent lower than the "practical" Fano factor of 0.13 which one ordinarily measures with germanium (as with silicon). It is not unreasonable to expect that silicon might follow suit in the future.

[6] See p. 57.

Under these two very optimistic conditions, the resolution of the Fe^{55} K-alpha spectrum might approach 100 eV, which is shown as the upper dashed line in Fig. 6. We cautiously suggest that this probably comes close to representing some sort of practical limit to what we might expect in system resolution at 5.9 keV.

Figure 7 illustrates the resolution with energy of silicon and germanium for the "practical" Fano factor of 0.13, and the "theoretical" Fano factor of 0.05. For comparison purposes, the resolution for NaI(Tl), proportional counters and gas ionization counters have been plotted also in Fig. 7, as they were in an earlier form of this illustration, which appeared in Ref *4*.

If we accept these as reasonable estimates, then we can conclude that one might hope to extend the resolution capabilities of silicon X-ray spec-

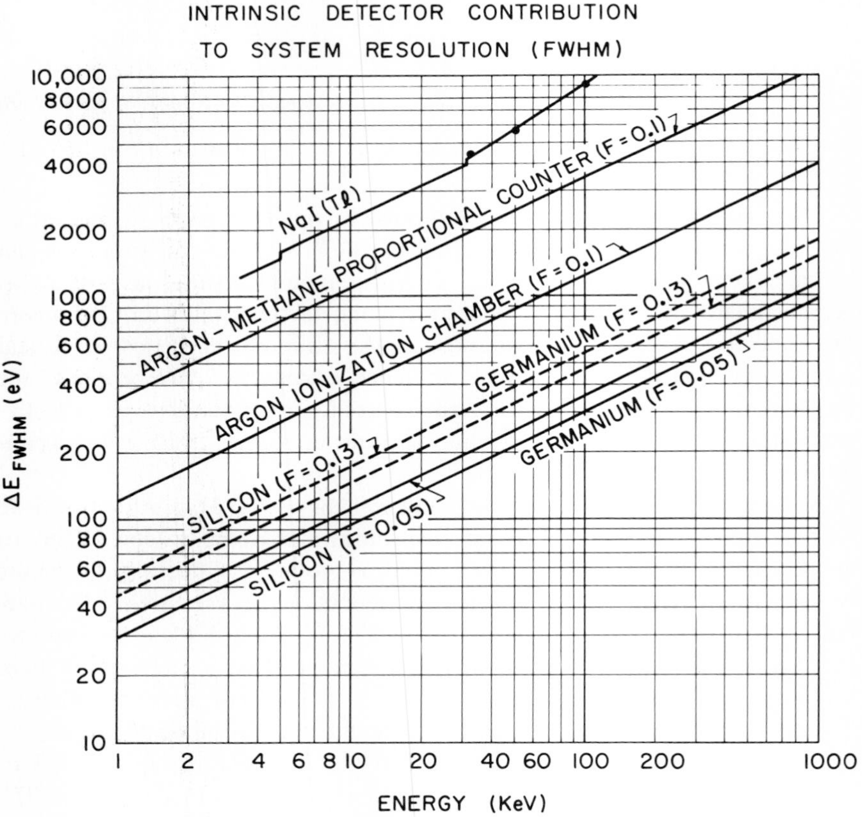

FIG. 7—*The resolution with energy of various types of energy-dispersive X-ray detectors. While the lower pair of lines for semiconductor detectors (F = 0.05) results from the recent work described in Ref 12, the upper pair of lines (F = 0.13) is a more useful approximation to what one can "realistically" expect with today's semiconductor detectors, exclusive of the contribution from electronic noise.*

trometers to just one lower position in the periodic scale, that is, to boron ($Z = 5$). In other words, all of the effort expended on behalf of improving the presently available resolution might gain us only one more element beyond where we should now be able to see, but it would contribute most certainly to an improvement in the accuracy and sensitivity of our quantitative analyses.

High resolution with very large detectors—While large area detectors comprise the subject of an entire paper currently being prepared [9], a few remarks here can be helpful to the present discussion.

Present detector technology permits the construction to commercial specifications of silicon detectors with areas up to at least 1250 mm^2. The resolution of a system with such large area detectors, though, is decidedly inferior to that of a system with small detectors. This is largely the fault of the electronics, rather than of the detector.

It is only with the very small silicon detectors (10 mm^2) that fundamental physical processes related to ionization statistics and charge collection fluctuations contribute in a significant way to the overall resolution of the system. Table 1 and Fig. 7 demonstrate (with the $F = 0.13$ values) that at the very low count rates, the electronic noise contributes about 35 percent of the total linewidth with the 10-mm^2 detector, about 75 percent with the 80-mm^2 detector, and about 83 percent with the 200-mm^2 detector.

The progressively greater electronic noise contribution with increasing detector area results largely from the "noise slope" of the field-effect transistor (FET) used in the first stage of the preamplifier, that is, the circuit noise increases with increasing capacitance at the FET gate. A reduction in the noise slope of the FET as a function of circuit input capacitance and of the baseline noise at short time constants would have an obviously beneficial effect on the resolution of the large-area X-ray systems. Such a reduction is well within the capability of present transistor technology.

The best we can do at the moment, therefore, is to express a word of optimism: the FET's probably will be improved in the future, and as they improve, so will the performance of X-ray spectrometers with large area detectors.

High resolution at high count rates—The count-rate capability of a circuit is related directly to the pulse-shaping time constant, but so is the resolution, generally in a conflicting manner. To maintain high resolution in the presence of a high count rate requires baseline restoration and pulse pileup rejection, to prevent linewidth broadening from pulse overlap, but it also requires clever preamplifier designs that permit the use of short time constants without a corresponding loss in resolution.

With the continuing innovative contributions by groups such as Goulding's at Berkeley [15,16], high-rate preamplifier designs can be expected to continue to improve in the future. The general acceptance of the larger area detectors will depend much on these improvements, too, so that the large detectors can be used in "normal" applications, as well as in those requiring high sensitivity, without having to sacrifice beam current or excitation source intensity in the interest of keeping the total count-rate low.

Toward Lower Energies—The full analytical capabilities of silicon X-ray spectrometers that are now available will not be realized until the detectors can be used either behind highly transparent but vacuum-tight window materials, or in a windowless or quasi-windowless configuration. Practical problems to both of these approaches are caused by the cold-trap action of the silicon detector and by the light sensitivity of semiconductor detectors. Still, for those applications in which the observation of very low energy X-rays is of principal concern, these limitations can be surmounted.

Absorption external to the detectors—Table 2 presents absorption calculations for the fluorescent K-shell X-ray spectra of sodium ($Z = 11$, 1041 eV), oxygen ($Z = 8$, 523 eV), and carbon ($Z = 6$, 282 eV) for different thicknesses of beryllium, and for two-centimeter paths in air and helium. The thicknesses selected for the beryllium calculations were based upon commonly used cryostat windows (0.002 and 0.001 in.) and for the thinner experimental windows (0.0005 and 0.0003 in.) now being tried on cryostats housing the very small silicon detectors. The two-centimeter path in air and helium was based upon the approximate distance between specimen and cryostat window, such as with the compact radioisotope source-target assembly shown in Fig. 8 [17]. Absorption coefficients were taken from the recent tables of Henke [18].

Several conclusions are immediately apparent from Table 2. For example, with the 2 and 1 mil beryllium windows, about one keV is the lower energy limit for useful quantitative analytical X-ray applications, and then only if the specimen is fluoresced and observed in a helium or vacuum environment. But with the 0.3 mil beryllium window and a vacuum or helium environment, analytical applications can be extended usefully down to about oxygen, although with not much sensitivity at the lower end of the energy spectrum.

Figure 9 illustrates X-ray transmission curves for two materials that have been found to be highly satisfactory in certain applications with gas-filled proportional counters at low X-ray energies [4], mylar, and formvar. Included also is the transmission of a 500 Å "light sealing" layer of aluminum.

While careful investigations demonstrated [19] that mylar of the thickness illustrated (0.00015 in.) is completely vacuum tight, it is known to be permeable to moisture. The unique feature of the mylar, though, is its fairly high transparency to the fluorescent radiations from oxygen and carbon, suggesting that it should be given serious consideration for those

TABLE 2—*Transmitted fraction of X-rays through beryllium, air, and helium.*

Element	$E(K\alpha)$,eV	Fraction Through Beryllium Windows				Fraction Through 2-cm Columns	
		0.002 in.	0.001 in.	0.0005 in.	0.0003 in.	Air	Helium
Sodium (Z = 11)...........	1041	0.007	0.05	0.29	0.47	4.3×10^{-4}	0.98
Oxygen (Z = 8)...........	523	5×10^{-17}	5.5×10^{-9}	8.3×10^{-5}	0.0036	4.0×10^{-18}	0.86
Carbon (Z = 6)...........	282	...	...	...	10^{-13}	4.8×10^{-5}	2×10^{-5}

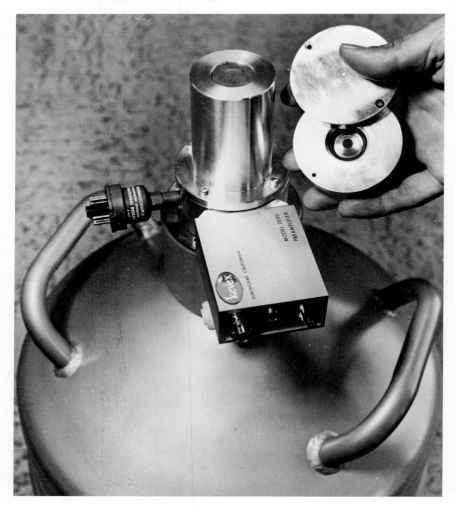

FIG. 8—*A compact radioisotope source-target assembly, adapted from the design by Giauque (Ref 19), allowing the specimen to be placed within two centimeters of the beryllium window of the cryostat.*

special applications in which the observation of oxygen and carbon X-radiation is important.

The formvar window described in Fig. 9 would be highly transparent down to and including the X-radiation from carbon, but with a very low efficiency for nitrogen, which lies between the carbon and oxygen absorption edges.

Absorption intrinsic to the detector—It is not of much use to extend the transmission of all windows external to the detector toward lower energies, if these same X-rays are stopped in the detector before they can interact in the sensitive region. Fortunately, as demonstrated in Table **3**, with presently available silicon detectors the absorption of X-rays in the detector gold contact and partially dead region is a very small effect in practical applications.

TABLE 3—*Transmitted fraction of X-rays through silicon and gold.*

Element	$E(K\alpha)$, eV	Fraction Through 0.5-μm Silicon	Fraction Through 0.1-μm Silicon	Fraction Through 200-Å Gold
Sodium ($Z = 11$)...	1041	0.82	0.96	0.90
Oxygen ($Z = 8$)....	523	0.30	0.79	0.82
Carbon ($Z = 6$)....	282	0.008	0.37	0.67

In Table 3, the X-ray transmission has been calculated for the same energies as in Table 2, for two different thicknesses of silicon and for one thickness of gold. Detector manufacturers have demonstrated that whereas the very best lithium-drifted silicon detectors occasionally can have partially dead layers with thicknesses as low as about 0.1 μm, most all of the good lithium-drifted silicon detectors have dead layers with thicknesses extending probably up to 0.5 μm. A typical gold surface-barrier contact is of the order of 200 Å thick. These three material thicknesses, therefore, have been chosen for presentation in Table 3.

It is seen in Table 3 that the gold contact causes no important absorption over the entire energy spectrum that is theoretically accessible to the spectrometer. It is also seen that even a 0.5-μm dead layer in silicon does not lead to serious absorption down to energies corresponding to the fluorescent X-radiation from oxygen, and that a 0.1-μm window permits satisfactorily high transmission all the way down to the fluorescent radiation from carbon.

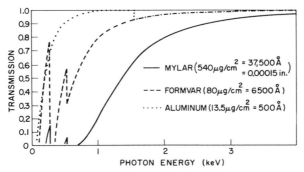

FIG. 9—*Calculated X-ray transmissions for thin layers of mylar; formvar, and aluminum. The discontinuities in the efficiency of mylar and formvar below 1 keV correspond to the absorption edges of oxygen (531 eV) and carbon (293 eV).*

These observations have several implications. First, although some workers are seeking lower-Z electrical contact materials to replace the gold, this is of no particular value to X-ray emission spectroscopy. This is not true, of course, regarding the use of these same detectors for the analysis of low energy electrons, for alpha particles or for other ions.

A second important conclusion is that the detectors with the ultra-thin dead layers are only significantly advantageous for X-ray observations at energies below about 500 eV. In X-ray emission spectroscopy this means only for the detection of carbon and nitrogen.

Finally, we should mention the potential development of ultra-thin window, low-noise silicon and germanium detectors by ion-implantation [20-24]. In these, junctions are formed by implanting boron and phosphorus ions on opposite sides of a small, thin wafer of high resistivity (greater than 10,000 ohm-cm) material and then overlaying thin surface-barrier electrical contacts on each side. Dead layers in ion-implanted silicon of less than 0.02 μm have been observed [21].

Again, though, reference to Table 3 suggests that the special qualities of ion-implanted detectors will not provide much advantage over the very best lithium-drifted silicon detectors in X-ray analyses all the way down to the practical limits of the electronic resolution. Ion-implantation will probably have a major impact on low energy electron and heavy particle and ion investigations, but will probably have a negligible impact on X-ray emission spectroscopy.

Unique Advantages of Silicon

We have attempted thus far to look into the future of silicon radiation detectors in X-ray applications without really justifying our concentration on silicon to the exclusion of germanium. In this final section, we review some of those attributes that make silicon uniquely advantageous for most X-ray analytical applications. We must carefully note, though, that only the last three items in the following material actually reflect physical differences inherent in the material. The first three are of technological origin, and most probably will be overcome in time.

Highest Attainable Resolution at Low Energies

It is apparent from Fig. 7 that in theory better resolution should be attained with germanium than with silicon. This results from the smaller value for "ϵ" in germanium (the average energy needed to produce an ion-pair), and from the apparent near-equality of the Fano factors in germanium and silicon [11,12]. This has not been realized in practice at the low energies, the reasons for which are not well understood.

The most remarkable accomplishments with germanium continue to come from Goulding's group at Berkeley where, as was mentioned earlier in this work, a *measured* upper limit of 0.075 for the Fano factor in germanium has been attained [*14*] at high energies. But at 8.04 keV (the fluorescent *K* X-radiation from copper) the same germanium detector yielded a FWHM of 175 eV, or slightly worse than the value that can be obtained with silicon, as illustrated in Fig. 3. This value corresponds to a "practical" Fano factor of about 0.13 at this energy. The resolution continued to deteriorate at still lower energies. The Berkeley group obtained no better than about 175-eV FWHM on the Fe^{55} spectrum as well.

This demonstrates that at the low energies (say below 8 keV or so) germanium is presently not even competitive on a resolution scale. The resolution at low energies with germanium undoubtedly can be improved, but in our opinion it will require some major solid-state surgery to accomplish this. Such surgery must be preceded by a better understanding of the causes of the poor behavior at low energies.

Uniform Production with Thin Windows

The attainment of 14-keV FWHM with 5.4-MeV alpha-particles in germanium [*25*] suggests that the Berkeley group made a germanium detector with a dead layer not thicker than about 0.1 μm. While this is fully adequate for most low energy applications (ignoring the resolution difficulties outlined above), the Berkeley group observed two disturbing features. First, the window thickness in germanium "grows" with time. The alpha resolution quoted above, for example, had deteriorated to 18 to 19 keV after a couple of days. And second, many detectors had to be manufactured before they happened to get the one with the very thin window.

There is apparently no method known at present to control the quality during manufacture to assure that an ultra-thin window will be realized in even a reasonable percentage of the time, or that it will not "thicken" in time. This suggests that production economics must certainly favor silicon.

May Be Thermally Cycled Without Damage

Illustrated in Fig. 10 are two compact configurations for the adaptation of silicon detectors to electron microprobes and scanning electron microscope (SEM) analyzers. (The left hand system, designed for an SEM, enables the detector to be placed close to the specimen for maximum X-ray collection efficiency.)

Apparent in Fig. 10 is the small size of the 8 to 10-h liquid nitrogen dewars that can be used with remote feed systems. A microprobe adapter, similar to that shown on the right in Fig. 10, is shown in Fig. 11, mounted in place on an Applied Research Laboratories (ARL) electron microprobe at the General Electric Laboratory at Vallecitos, Calif. The automatic liquid nitrogen feeder extends from the dewar on the floor.

The silicon detector systems shown in Fig. 10 can be mounted warm upon the SEM or electron microprobe assembly, and then cooled down for

FIG. 10—*Two silicon X-ray spectrometers adapted for use on SEM* (left) *and electron microprobe* (right) *systems, showing the small dewar size that can be used with remote liquid nitrogen feed.*

use. If they inadvertently warm during dewar change, in the event that the master dewar is depleted, or in the event of a malfunction with the automatic nitrogen feeder, silicon detectors suffer no damage. This amounts to a built-in "insurance" factor against financial loss and experiment "down" time. In fact, the highest resolution silicon detectors are still rated for 25 thermal cycles up to room temperature and back down to liquid nitrogen without suffering any observable degradation in performance. Such a compact assembly could not be as safely utilized with lithium-drifted germanium detectors, since the germanium detectors are ruined essentially in just one warming.

Maximum Analytical Convenience

The first two advantages for silicon discussed above centered on the very low energies, below 10 keV or so. But a major physical limitation with germanium causes silicon to be favored for analytical applications extend-

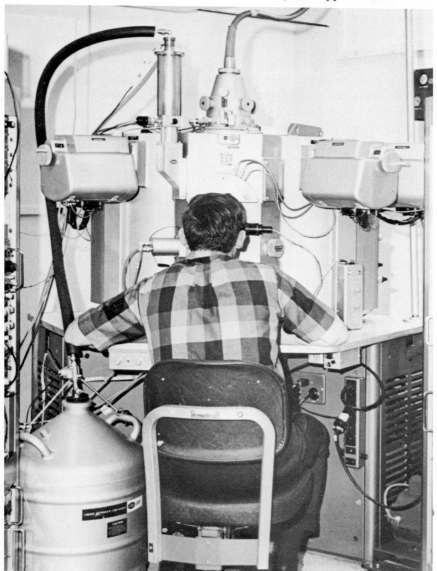

FIG. 11—*The right hand cryostat illustrated in Fig. 10 is shown mounted in place on an ARL electron microprobe at the General Electric Laboratory in Vallecitos, Calif. The automatic liquid nitrogen feeder extends from the dewar on the floor.*

ing up to about 40 keV. This is the nonlinear efficiency response in germanium that results from the combined actions of the absorption discontinuity across the K-absorption edge at 11.103 keV, and the relatively high escape fraction in germanium. The magnitude of this nonlinearity was calculated by Palms et al [26], in terms of absolute photopeak efficiency, and by Walter [1] in a very instructive illustration of the full energy peak efficiency for both germanium and silicon detectors.

The results of these analyses demonstrate that germanium begins to drop in efficiency at about 40 keV, reaching a value of only 50 percent just above the K-absorption edge, and then making a discontinuous jump in efficiency by a factor of two at the absorption edge. This kind of behavior, which differs from detector to detector according to the magnitude of the dead layer, and which may also vary in time as the dead layer "grows," can cause serious practical difficulties in the quantitative interpretation of X-ray spectra observed with a germanium detector.

The efficiency of silicon also varies, but there are no significant variations in the efficiency function for silicon above a few keV, and the corresponding efficiency discontinuity in silicon at its own K-absorption edge of 1.838 keV is less than that for germanium by more than a factor of ten. One can therefore conclude that below about 40 keV silicon detectors probably simplify quantitative chemical analyses in X-ray emission studies. The K-series X-ray spectra for 65 percent of the naturally occurring elements fall below this energy, while the L and higher order series of X-ray spectra for *all* presently identified elements fall well below this value.

Most Suitable for X-ray Energies Common to Scanning Electron Microscopy

Most scanning electron microscopes (SEM) and electron microprobes are operated at accelerating potentials of 50 kV or less, suggesting that the X-rays from the specimen are also produced with energies of less than 50 keV, and in significant numbers with energies below 40 keV. This falls precisely within the range in which silicon is more convenient for analytical applications, as discussed above. SEM applications also can extend down to very low energies because of the vacuum environment, again favoring silicon detectors.

Most Suitable for Quantitative Chemical Analyses

It is well known that the maximum accuracy in quantitative X-ray fluorescence analysis is obtained at the lower energies, where the background is minimized and where energy losses through Compton scattering are reduced. This makes it desirable to utilize the L X-ray spectra from the heavier elements for quantitative analysis, rather than the more energetic K X-rays. The L X-ray spectra for all of the elements fall below 20 keV, making them uniquely suitable for analysis with silicon detectors.

Finally, it is important for the system to be as insensitive as possible to background at higher energies, which can reduce the sensitivity at the lower energies. Again, this is far better satisfied with silicon than with germanium.

References

[1] Walter, F. J., "Proceedings of the 12th Scintillation and Semiconductor Counter Symposium," Washington, D. C., 1970, to be published in *Transactions on Nuclear Science,* Institute of Electrical and Electronic Engineers, Vol. NS-17, No. 3, 1970.
[2] Frankel, R. S. and Aitken, D. W., to be published in *Applied Spectroscopy,* Sept.-Oct. 1970.
[3] Burkhalter, P. G. and Campbell, W. J., "Proceedings of the Second Symposium on Low Energy X and Gamma-Sources and Applications," Austin, Tex., 1967, Technical Report ORNL-IIC-10.
[4] Aitken, D. W., *Transactions on Nuclear Science,* Institute of Electrical and Electronic Engineers, Vol. NS-15, No. 3, 1968, p. 10.
[5] Campbell, W. J., Burkhalter, P. G., and Marr, M. E., III, 8th National Meeting Society for Applied Spectroscopy, Anaheim, Calif., Oct. 1969. Abstract in *Applied Spectroscopy,* Vol. 23, No. 6, 1969, p. 670.
[6] Zulliger, H. R. and Aitken, D. W., *Transactions on Nuclear Science,* Institute of Electrical and Electronic Engineers, Vol. NS-14, No. 1, 1967, p. 563.
[7] Zulliger, H. R. and Aitken, D. W., *Transactions on Nuclear Science,* Institute of Electrical and Electronic Engineers, Vol. NS-15, No. 1, 1968, p. 466.
[8] Zulliger, H. R., Middleman, L. M., and Aitken, D. W., *Transactions on Nuclear Science,* Institute of Electrical and Electronic Engineers, Vol. NS-16, No. 1, 1969, p. 47.
[9] Frankel, R. S. and Aitken, D. W., paper to be presented at the "Third Symposium on Low Energy X and Gamma-Sources and Applications," Boston, Mass., 10 June 1970, and to be published in the *Symposium Proceedings.*
[10] Elad, E., Sandborg, A. O., Russ, J. C., and Van Gorp, T., *Transactions on Nuclear Science,* Institute of Electrical and Electronic Engineers, Vol. NS-17, No. 1, 1970, p. 354.
[11] Zulliger, H. R. and Aitken, D. W., "Proceedings of the 12th Scintillation and Semiconductor Counter Symposium," Washington, D. C., March, 1970, to be published in *Transactions on Nuclear Science,* Institute of Electrical and Electronic Engineers, Vol. NS-17, No. 3, 1970.
[12] H. R. Zulliger, to be submitted to the *Physical Review.*
[13] Radeka, V., "Proceedings of the 12th Scintillation and Semiconductor Counter Symposium," Washington, D. C., March 1970, to be published in *Transactions on Nuclear Science,* Institute of Electrical and Electronic Engineers, Vol. NS-17, No. 3, 1970.
[14] Pehl, R. H. and Goulding, F. S., Lawrence Radiation Laboratory preprint No. UCRL 19439, Berkeley, Calif., 1970. To be published in *Nuclear Instruments and Methods,* 1970. While the preprint describes measurements which yield $F = 0.08$ in germanium, a value of $F = 0.075$ has been subsequently measured by the same group (R. H. Pehl, private communication.)
[15] Goulding, F. S., Walton, Jack, and Malone, D. F., *Nuclear Instruments and Methods,* Vol. 71, 1969, p. 273.
[16] Goulding, F. S., Walton, Jack, and Pehl, R. H., *Transactions on Nuclear Science,* Institute of Electrical and Electronic Engineers, Vol. NS-17, No. 1, 1970, p. 218.
[17] Giauque, R. D., *Analytical Chemistry,* Vol. 40, Nov. 1968, p. 2075.
[18] Henke, B. L., *Norelco Reporter,* Vol. XIV, 1967, p. 75.
[19] Culhane, J. L., Herring, J., Sanford, P. W., O'Shea, G., and Phillips, R. D., *Journal of Scientific Instruments,* Vol. 43, 1966, p. 908.

[*20*] Meyer, O., *Transactions on Nuclear Science,* Institute of Electrical and Electronic Engineers, Vol. NS-15, No. 3, 1968, p. 232.

[*21*] Meyer, O., *Nuclear Instruments and Methods,* Vol. 70, 1969, p. 279.

[*22*] Meyer, O., *Nuclear Instruments and Methods,* Vol. 70, 1969, p. 285.

[*23*] Eriksson, L., Davies, J. A., and Mayer, J. W., *Science,* Vol. 163, No. 3868, 1969, p. 627.

[*24*] Sebillotte, Ph., Siffert, P., and Coche, A., *Transactions on Nuclear Science,* Institute of Electrical and Electronic Engineers, Vol. NS-17, No. 1, 1970, p. 24.

[*25*] Pehl, R. H., Goulding, F. S., and Landis, D. A., *Nuclear Instruments and Methods,* Vol. 59, 1968, p. 45.

[*26*] Palms, J. M., Rao, P. Venugopala, and Wood, R. E., *Transactions on Nuclear Science,* Institute of Electrical and Electronic Engineers, Vol. NS-16, No. 1, 1969, p. 36.

Emanuel Elad[1]

Low-Noise Cryogenic Preamplifiers

REFERENCE: Elad, Emanuel, **"Low-Noise Cryogenic Preamplifiers,"** *Energy Dispersion X-ray Analysis: X-ray and Electron Probe Analysis, ASTM STP 485,* American Society for Testing and Materials, 1971, pp. 57–81.

ABSTRACT: Low-noise cryogenic preamplifiers for semiconductor X-ray detectors will be described. The basic structure of field-effect transistor (FET) preamplifier and its major parameters in the context of the X-ray spectrometer will be presented. Operational characteristics of an FET and its noise properties, especially at cryogenic temperatures, will be outlined. Noise frequency response requiring band-pass limitations and their achievement through pulse shaping in the main amplifier will be described. Noise formulation will be presented emphasizing the influence of various parameters and the conflicting requirements of low-noise and high count-rate performance. Various feedback schemes will be described, comparing in particular the noise and count-rate characteristics of the opto-electronic and the resistive feedback configurations. Recent experimental results with silicon and germanium spectrometers will be given and compared with the theoretical limitations. Finally, future possibilities in the development of X-ray energy analyzers will be assessed.

KEY WORDS: semiconductor devices, spectrometers, silicon, germanium, X-ray spectra, X-ray analysis, electromagnetic noise, cryogenics, preamplifiers, solid state counters, dispersing, energy methods

The main impetus in the rapid growth and application of the X-ray energy analyzing technique was given by the development of the low-noise cryogenic preamplifier [1,2].[2] The energy analysis method has the advantages of high efficiency and short processing time over the wavelength dispersive method and various nondispersive methods using selective filters or differential adsorption for energy discrimination [3,4]. Its main drawback in the early days, when it had to be implemented with gaseous or scintillation detectors and vacuum tube preamplifiers, was its poor energy resolution (> 2 keV). As this quantity represents the ability of the system to distinguish between adjacent lines in the spectrum we see from Fig. 1 that such systems with 2-keV resolution were not very useful. However, the use of cooled semiconductor detectors in conjunction with cryogenic

---

[1] Senior physicist, ORTEC, Inc., Oak Ridge, Tenn. 37830.
[2] The italic numbers in brackets refer to the list of references appended to this paper.

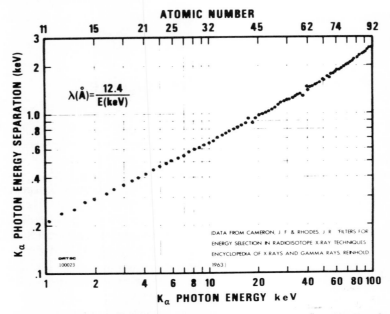

FIG. 1—*Spectral line separation as a function of energy and atomic number.*

preamplifiers using field-effect transistors (FET's) [2,5] has changed the situation considerably. Resolutions of 500 eV (Si) full width at half maximum (FWHM) became available, and the technique proved useful for higher Z elements [6]. Consequent improvements in the preamplifier during the last four years brought the electronic resolution down to about 80 eV (Si), with system resolutions of 160 eV for 6-keV X-rays commercially available. Therefore, nowadays the noise of limitation on the detection of low Z elements was extended down to carbon ($Z = 6$, $E_\alpha = 283$ eV). However, before routine commercial usefulness of the energy analysis technique may be extended to that low Z limit, window problems in the spectrometer will have to be solved.

This paper will describe the structure of a cryogenic FET preamplifier and its major parameters. Operational characteristics of an FET and its noise properties, especially at cryogenic temperature, will be outlined. Noise frequency response requiring band-pass limitations and their achievement through pulse shaping in the main amplifier will be described. Noise formulation will be presented emphasizing the influence of various parameters and the conflicting requirements of low-noise and high count-rate performance. Various feedback schemes will be described, comparing in particular the noise and count-rate characteristics of the opto-electronic and the resistive feedback configurations. Recent experimental results with silicon and germanium spectrometers will be given and compared with the theoretical limitations. Finally, future possibilities in the development of X-ray energy analyzers will be assessed.

General

A block diagram of a semiconductor X-ray spectrometer is given in Fig. 2. Only the most essential modules of the system are included, as the main purpose of Fig. 2 is to illustrate the relative position of the preamplifier in the X-ray spectrometer.

The X-rays striking the detector generate electron-hole pairs which when swept away by the electric field produce a current impulse at the input of the preamplifier. The total number of electron-hole pairs and consequently the integrated current is proportional to the energy of the X-ray. The width of the impulse is the transit time of the charges in the detector and thus depends on detector thickness. The current impulse is integrated by the preamplifier producing at its output a voltage pulse with rise time in the range of 50 to 100 nanoseconds and fall times of 50 to 1000 μs. That pulse is further amplified and shaped in the amplifier resulting in a voltage pulse of few volts and rise time of 2 to 20 μs. The output pulse of the amplifier may be symmetric or may have a longer fall time depending on the type of shaping used. Subsequent to amplification the pulses are sorted in the multichannel pulse-height analyzer and the energy spectrum recorded.

This paper will describe mainly the preamplifier module, although some characteristics of the amplifier connected with improving the signal-to-noise ratio of the system will be detailed. The main role of the preamplifier is to

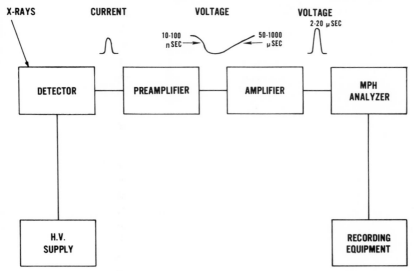

FIG. 2—*A block diagram of a semiconductor X-ray analyzer.*

amplify the output pulses of the detector with minimal deterioration of their signal-to-noise ratio. To fully appreciate that statement, let us consider the signals which are processed by the preamplifier. The charge generated in the detector by an X-ray with energy E is given by

$$\Delta Q = (E/\epsilon)q \dots\dots\dots\dots\dots\dots\dots\dots(1)$$

where ϵ is the average energy necessary to produce one electron-hole pair and q is the charge of an electron.

$$\epsilon \text{ (Si; 77 K)} = 3.8 \text{ eV}$$

$$\epsilon \text{ (Ge; 77 K)} = 2.98 \text{ eV}$$

$$q = 1.6 \ 10^{-19} \text{ Coul}$$

The amplitude of the current impulse at the input of the preamplifier is given by

$$I = \Delta Q/\Delta t \dots\dots\dots\dots\dots\dots\dots\dots(2)$$

where Δt is the transit time of the detector. In case of the planar type detector Δt equals

$$\Delta t = d/v = d^2/\mu V \dots\dots\dots\dots\dots\dots(3)$$

where:

  $d =$ thickness of the detector,
  $v =$ velocity of the carriers,
  $\mu =$ mobility of the semiconductor, and
  $V =$ voltage across the detector.

For silicon and germanium at 77 K carrier velocity saturates for fields above 10^3 V/cm which are normally attained in X-ray semiconductor detectors. Therefore, combining Eqs 1, 2, and 3 gives

$$I = Eqv_{sat}/\epsilon d \dots\dots\dots\dots\dots\dots\dots\dots(4)$$

where $v_{sat} \cong 10^7$ cm/s.

For example, for a 6-keV X-ray the current impulse from a silicon detector 4 mm deep will be 6.3 nanoamperes. If one compares this current with typical leakage current of silicon p-n junctions at room temperature, and especially that of large area diodes (typically 0.5 μA), it is obvious why cryogenic operation of the front section of the spectrometer is essential.

  In the front section of the preamplifier (within the charge-sensitive loop) the signal level will be

$$V = \Delta Q/C_f \dots\dots\dots\dots\dots\dots\dots\dots(5)$$

where C_f is the feedback capacitance of the loop. Thus for the 6-keV X-ray with $C_f = 1$ pF the output pulse of the charge-sensitive section will be

0.25 mv. To preserve the information in these minute pulses submicrovolt noise is essential in the front section of the amplifying chain.

Let us now summarize the important characteristics of the preamplifier module.

(a) Low noise is necessarily the major feature of the preamplifier (as explained above).

(b) Count-rate capability: The resolution of the preamplifier should have negligible dependence on count rate over a large dynamic range of rates. That characteristic is particularly important in quantitative analysis, where variations of resolution with changing count rate will complicate the unfolding of the spectrum. As will be explained later the requirement for high count-rate capability is inconsistent with optimum low noise performance at low count rates.

(c) Linearity and dynamic range: The transfer function of the preamplifier should be linear over a wide range of voltage (normally 10 V). This characteristic again determines the extent of usefulness of quantitative analysis by the energy analysis method.

(d) Stability: The gain of the preamplifier should be very stable with temperature variations or power supply fluctuations. Any instability will have the effect of higher noise, spreading the widths of the measured lines.

The Preamplifier

Two types of preamplifiers are available: voltage-sensitive and charge-sensitive. In the voltage-sensitive preamplifier the output voltage is proportional to the input voltage, while in the charge-sensitive circuit the output is proportional to the input charge. In the case of semiconductor radiation detectors which are capacitive transducers the charge-sensitive preamplifier is preferred, because its gain and gain stability do not depend on the capacitance of the detector and the stabiliy of high gain d-c amplifier. However, the noise of a voltage-sensitive configuration is slightly smaller, and therefore it may be attractive for very high resolution systems using small detectors. The reader interested in more information on the voltage-sensitive configuration may consult Refs 7, 8, and 9. Below, only the characteristics of the more popular charge-sensitive preamplifier will be described.

A block diagram of a cryogenic charge-sensitive preamplifier is given in Fig. 3. The preamplifier can be divided into two parts: the input FET stage, which together with the feedback elements and the semiconductor detector is enclosed in the cryogenic chamber, and the rest of the amplifying stages which are operated at room temperature. The noise properties of the preamplifier are determined mainly by its cryogenic part, while its other characteristics are determined by various sections of the instrument. The first amplifying section A_1 together with the cryogenic part comprise the charge-sensitive loop which integrates the input charge on C_f (Eq 5). The open loop gain of the charge-sensitive section (A) must be large to

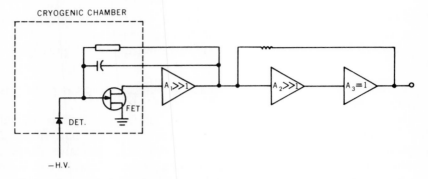

FIG. 3—*A block diagram of a cryogenic FET preamplifier.*

provide C_f $(1-A)$ $\gg C_{in}$ ($C_{in} =$ input capacitance of the preamplifier) and thus ensure the validity of Eq 5. The d-c feedback around the charge-sensitive loop provides a path for the leakage currents of the detector and the FET and the average signal current from the detector. The output voltage pulse from the charge-sensitive loop is amplified by a wide-band feedback amplifier A_2. This is normally an operational amplifier, and its gain is determined by passive elements. The pole-zero cancellation network, when used, is located between the amplifiers A_1 and A_2, and its purpose is to shorten the decay time of the output pulse of the charge-sensitive loop. This will decrease the probability of pileup thereby improving the count-rate capabilities of the preamplifier. The output stage A_3 provides low output impedance for driving coaxial cables. It has a gain of unity and usually determines the linearity of the circuit. Let us now discuss in more detail the cryogenic part of the preamplifier.

The Field-Effect Transistor (FET)

The heart of the input stage is the FET device. Figure 4a shows a cross section of a junction FET, which is the lowest noise device in its class. When voltage is applied between the drain and the source, current flows through the *n*-channel. This current may be modulated by reverse-biasing the gate-to-channel junction and thus narrowing the undepleted cross section of the channel. The input impedance of the device (into the gate) is very high (reverse-biased *p-n* junction), and therefore its low-noise qualities are especially valuable for high impedance transducers such as the semiconductor radiation detector. Figure 4b shows the current-voltage characteristics of an FET with the gate voltage as a parameter. We see that above a certain drain voltage (pinch-off voltage) the drain current is independent of the voltage and decreases with the gate-to-source bias. For low noise operation the FET is operated in the saturation region. As will be

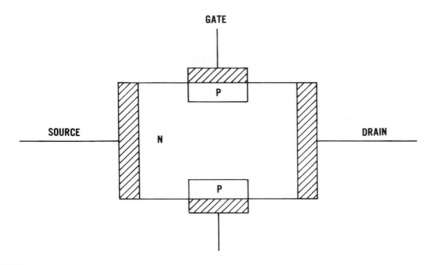

FIG. 4a—*Cross-section of a junction field-effect transistor.*

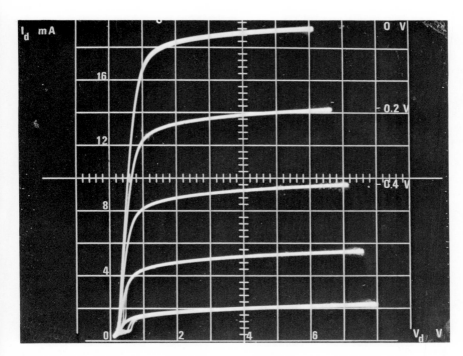

FIG. 4b—*Current-voltage characteristics of a junction FET (gate voltage as a parameter).*

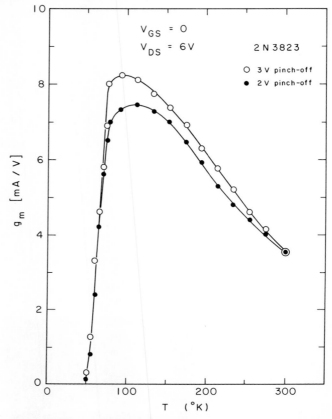

FIG. 5—*Transconductance versus temperature of a silicon FET.*

clarified below an important figure of merit of an FET for low-noise operation is the ratio g_m/C_{in} where g_m represents the gain of the device (transconductance) and C_{in} its input capacitance. That ratio equals the maximum frequency of operation of an FET, and therefore very high frequency
(VHF) transistors are preferred for high resolution preamplifiers. The
two main noise sources of an FET are the thermal noise of the channel and
the leakage current of the gate junction. Intuitively, it is clear that the
noise of the FET will decrease with temperature as both noise sources are
thermal in origin. The important parameter of the input stage is its signal-
to-noise ratio, and thus the question that remains confronting the cryogenic operation of FET's is what happens to the gain (g_m) at low temperatures. Figure 5 shows a typical curve of g_m versus temperature for a silicon
FET. The transconductance increases with decreasing temperature down
to a certain temperature below which it decreases rapidly. FET's from
other materials such as germanium or gallium arsenide will operate down

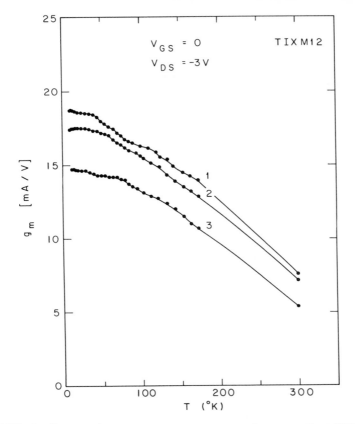

FIG. 6—*Transconductance versus temperature of a germanium FET.*

to liquid helium temperatures as demonstrated by Fig. 6 for a germanium device. The optimum temperature for silicon FET's is around liquid nitrogen temperature, and it varies with the type of the FET and from device to device of the same type.

The characteristics of field-effect transistors have been described very briefly. The interested reader may find additional general information in Refs 9 and 10. More about cryogenic operation of FET's will be found in Refs 5, 8, and 11.

Input Stage Configurations

The performance of a very high resolution preamplifier is determined not only by the FET, but also by the detector-preamplifier coupling network, the feedback elements, and microphonic noise generation in the system. Two basic methods for coupling the detector and the preamplifier (a-c and d-c coupling) are illustrated in Fig. 7. The a-c coupling tech-

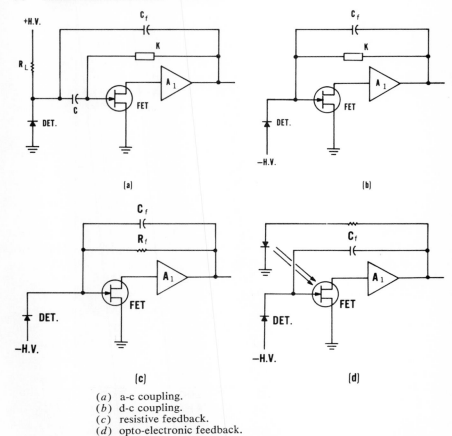

(a) a-c coupling.
(b) d-c coupling.
(c) resistive feedback.
(d) opto-electronic feedback.

FIG. 7—*Detector-preamplifier coupling networks and feedback schemes.*

nique, Fig. 7a, provides convenience in detector mounting (one side may be grounded), but is inherently noisier and therefore generally not used in high resolution preamplifiers. In the d-c coupled configuration, Fig. 7b, the following noise sources are eliminated: the bias resistor R_L and the high voltage coupling capacitor C. Also, microphonic generation is reduced by removing the high voltage from the input of the preamplifier. However, in this configuration the detector has to be electrically insulated from ground by a good heat conductor.

The feedback elements C_f and K also add noise to the system, but are necessary to maintain vital functions of that system. The a-c feedback through C_f ensures the charge-sensitivity, which offers high gain stability and independence of the gain from the size of the detector (that is detector capacitance). The value of C_f may be reduced by increasing A_1 (Fig. 3), and the modern cryogenic preamplifiers for X-ray detectors use capacitors of less than 1 pF. The feedback element K which provides the return path for d-c input currents, both quiescent and signal average, presents a more serious problem because of its higher noise. At the present there are two main approaches for realization of K, although a third theoretically promising method has been proposed. The simplest approach is to use resistive feedback, Fig. 7c. High megohm resistors are necessary (see section on noise calculations), and these generally suffer from irregular dependence on frequency [12] and add some stray capacitance to the input. The resistor also determines the energy rate product capability of the preamplifier. Increasing the resistor decreases its noise but also decreases the energy rate capability of the system. Here we see the conflicting requirements on the resistor for low-noise and high count-rate performance.

The second approach for realization of K is the opto-electronic feedback [13,14], Fig. 7d, in which the leakage currents of the FET are varied by light emitted by a GaP diode powered by the output voltage of the preamplifier. In this approach the resistor is eliminated, but the input leakage current, which introduces the same type of noise as the resistor, is almost doubled because the FET gate current is increased to the level of detector current by photoelectric generation. In some cases (depending on the leakage current of the detector) the resulting noise will be higher than for the case of resistive feedback. Moreover, in the case of the opto-electronic feedback the noise is count-rate dependent because the mean current through the detector is proportional to the rate (see section on noise calculations). This results in increased leakage currents of the FET and consequently a variation of resolution width with count rate. That problem may be particularly bothersome when a complex spectrum obained by electron excitation (varying count rate) has to be unfolded and quantitatively analyzed. Resolution (for manganese X-rays) versus count-rate performance of the resistive and opto-electronic feedback configurations is compared in Fig. 8. We see that for both sets of peaking times the resolution of the opto-electronic feedback configuration deteriorates rapidly above 1000 counts per second. Part of that deterioration is pulse pileup (also above 5000 counts per second (cps) for resistive feedback 18 μs peaking time–50 μs pulse width), which increases strongly with pulse width. However, for the optical feedback a large part of resolution degradation at higher count rates is due to noise increase in the feedback network [13] (see section on noise calculations). The noise of the resistive feedback configuration is to the first order independent of count rate, making it superior to the opto-electronic feedback configuration, especially at high count rates.

A third feedback scheme that theoretically should introduce no noise into the preamplifier is pulsed feedback. The idea here is to remove the charge that generates the input currents (quiescent and signal average) in pulses instead of continuously through the feedback resistor or the FET (optical feedback). Any input pulses which might be distorted because

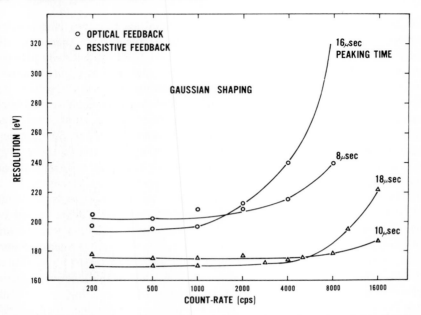

FIG. 8—*Resolution versus count rate for opto-electronic and resistive feedback configurations. Semilogarithmic scale.*

they arise during the discharge pulse can be rejected. In between the discharge pulses, the preamplifier should be free from noise generated in the feedback network. The pulsed feedback method which was first proposed by Kandiah and Stirling [14] employed an opto-electronic scheme. Recently another variation of the pulsed feedback method was proposed by Radeka [15] in which the input charge is removed by forward biasing the gate-to-channel junction of the input FET. As of now the experimental results obtained with the pulsed feedback method have not equaled the performance capabilities of the two previously described techniques. Any pulsed feedback method will require a more complicated and expensive amplifying chain that is needed for a continuous feedback configuration.

Microphonic generation may be a potentially significant noise source. In every high impedance capacitance transducer, a charge or voltage signal will be generated with variation of the capacitance according to

$$dQ = V \; dC \dots \dots \dots \dots \dots (6)$$

In cryogenic systems mechanical vibrations inside the cryostat are induced by bubbling of liquid nitrogen and by environmental noise. Vibrations of the input circuit components with respect to surfaces at different potentials will cause variations in the input capacitance and subsequently generate a noise signal dQ. The gate of the FET is held at a potential of a few tenths of a volt, while surfaces of several hundred volts (detector bias), few volts (drain of the FET), and ground (cryostat) are present in its vicinity. Normally, the microphonics are induced at the gate of the FET by the high voltage surfaces, although other combinations have been observed. To realize the seriousness of the problem, we observe (see Eq 6) that to generate a microphonic signal with the amplitude of 10 keV (Si) at the input of the preamplifier (detector bias 1000 V) a change of capacitance of only 5×10^{-7} pF is needed. Fortunately the frequency of the mechanical vibrations is inherently small, normally below 1 kHz, and therefore a large portion of the induced noise will be filtered out in the amplifier (see next section). Microphonic generation is strongly dependent on the housing arrangement of the cryogenic part of the preamplifier. An example of experimental cryostat is shown in Fig. 9. The input stage of the preamplifier is mounted on the cold surface. The detector which is d-c coupled to the FET is insulated from ground by the two ceramic rods having good thermal conductivity. Various cryostats are necessary to interface the X-ray spectrometer with the different types of electron microscopes and microprobes. Obviously, because of their different construction these cryostats have different microphonic properties.

The main characteristics of low-noise cryogenic preamplifiers have been described. Additional information on the subject may be found in Refs 5, 12, 16, and 17.

The Amplifier

The output pulses from the preamplifier are further processed in the main amplifier to obtain the amplitude and shape required by the pulse-height analyzer. In this section only those aspects of the amplifier which affect the signal-to-noise ratio of the system will be described.

The frequency response of the noise output of the preamplifier is shown in Fig. 10a. We see that the noise is high at very low and very high frequencies, but has a plateau in the intermediate frequency range. The frequency spectrum of the signal (see Fig. 2 for waveform) on the other hand is completely different with the largest components at the intermediate frequencies, Fig. 10b. It is clear therefore that filtering the output of the preamplifier will improve the signal-to-noise ratio significantly. The response of an ideal filter is shown in Fig. 10c. Practical filters can only approach such response with the optimum arrangement depending strongly on the particular noise spectrum of interest. Resistor-capacitor (RC) networks provide the simplest form of frequency filters (see Fig. 11). The differentiator, Fig. 11a, is a high-pass filter and the integrator, Fig. 11b, a low-pass filter. The filters are characterized by their corner frequencies f_1 and f_2 at which the gain decreases to 0.707 of the band-pass value. The corner frequencies are inversely proportional to the respective time constants. Long time constants represent low frequency band-pass and consequently slow rising and falling (wide) output pulses. Combining an RC differentiator

FIG. 9—*The housing of the cryogenic part of the X-ray analyzer.*

and integrator with buffer amplifying stages in between, Fig. 11*c*, gives a band-pass filter of the type required for the improvement of the signal-to-noise ratio, Fig. 10*d*. Similar configurations of differentiators and integrators with variable time constants are built into the main amplifier. A popular band-pass filter used in most modern amplifiers is the Gaussian shaper which provides a better signal-to-noise ratio than the simple passive RC circuit. The true Gaussian requires a single differentiator and an infinite number of integrators. In practice Gaussian shapers usually have only two to four integrators and are therefore really only semi-Gaussian.

Extensive work has been and is being done on optimization of filters for high resolution nuclear spectroscopy. The interested reader may consult Refs *18, 19,* and *20.*

Noise Calculations and Results

Previous paragraphs dealt with the structure and performance characteristics of the cryogenic preamplifier. In this section some quantitative information on noise properties will be given in an attempt to classify the
relative importance of the various noise sources. Also, typical results obtained with silicon and germanium X-ray spectrometers will be presented
and compared with theoretical capabilities of the detector.

The mean square noise voltage at the output of a charge-sensitive preamplifier with resistive feedback is given by [16]

$$V_{no}^2 = 4kT \left\{ \frac{0.6}{g_m} \frac{(C_{FET} + C_{det} + C_f)^2}{C_f^2} + \frac{1}{\omega^2 C_f^2} \left[\frac{1}{R_f} + \frac{q(I_{det} + I_{FET})}{2kT} \right] \right\} \quad (7)$$

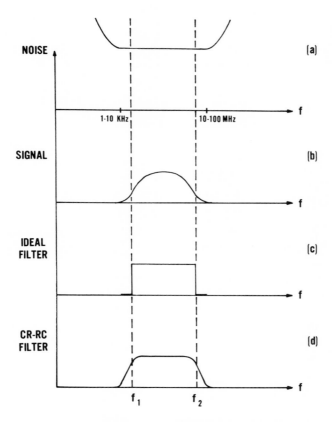

(*a*) Noise of the preamplifier versus frequency.
(*b*) Signal frequency response.
(*c*) Frequency response of an ideal filter.
(*d*) Frequency response of a CR-RC filter.
FIG. 10—*Band-pass limitations for the improvement of the signal-to-noise ratio.*

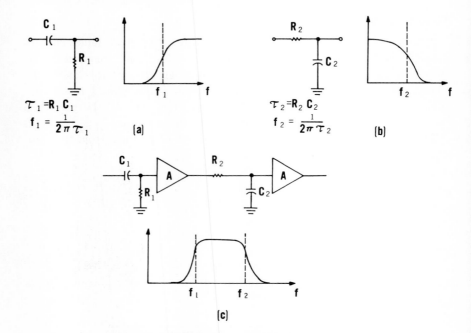

(a)   RC differentiator and its frequency response.
(b)   RC integrator and its frequency response.
(c)   RC differentiator and integrator forming a band-pass filter.
      f_1 and f_2 are the 3 db corner frequencies.

FIG. 11—*Passive RC filters.*

where:

$$k = \text{Boltzman constant } (k = 1.38 \times 10^{-23} \text{ joules/deg K}),$$
$$T = \text{absolute temperature,}$$
$$\omega = 2\pi f, \text{ and}$$

$I_{\text{det}}, I_{\text{FET}} = $ leakage currents of the detector and the FET, respectively. The first term in Eq 7 is the thermal noise of the FET (in series with the input—series noise), and the second term the noise from the feedback resistor and the leakage currents (in parallel with the input—parallel noise). For the case of the opto-electronic feedback, when $I_{\text{det}} > I_{\text{FET}}$, the average noise voltage is given by

$$V_{\text{no}}^2 = 4kT\left\{\frac{0.6(C_{\text{FET}} + C_{\text{det}} + C_f)^2}{g_m C_f^2} + \frac{1}{\omega^2 C_f^2}\left[\frac{q(2I_{\text{det}} + I_s)}{2kT}\right]\right\}\dots(8)$$

where I_s is the average signal current,

$$I_s = (E/\epsilon)qN\dots\dots\dots\dots\dots\dots\dots(9)$$

and N is the count rate in counts per second.

For the case when $I_{\text{FET}} > I_{\text{det}}$, an additional light-emitting diode must be pointed toward the detector to increase its leakage to the level of I_{FET}. The first light-emitting diode (see Fig. 7d) will still serve as the feedback element. For this case the term $2\,I_{\text{det}}$ in Eq 8 should be replaced by $2\,I_{\text{FET}}$.

The equivalent noise charge (ENC) is a convenient way to express the noise-to-signal ratio of a charge-sensitive preamplifier [16]. Using CR-RC filters with equal time constants (τ) the ENC for the filtered noise of the resistive feedback configuration is given by

$$\text{ENC}^2 = \frac{e^2}{2}\left\{\frac{0.6kT}{\tau g_m}(C_{\text{FET}} + C_{\text{det}} + C_f)^2\right.$$

$$\left. + kT\tau\left[\frac{1}{R_f} + \frac{q}{2kT}(I_{\text{det}} + I_{\text{FET}})\right]\right\}\quad\ldots(10)$$

$$e = 2.712$$

when divided by q the ENC is expressed in electrons, root mean square.

The more widely used measure of noise is the resolution expressed in full width at half maximum (FWHM) of a Gaussian peak. The resolution given in units of energy must be normalized by ϵ of the specific detector material. The noise in terms of FWHM is given by

$$\text{FWHM (eV)}_{\text{electronic}} = \frac{2.35\epsilon}{q}\,\text{ENC}$$

$$= \frac{4.52\epsilon}{q}\left\{\frac{0.6kT}{\tau g_m}(C_{\text{det}} + C_{\text{FET}} + C_f)^2\right.$$

$$\left. + kT\tau\left[\frac{1}{R_f} + \frac{q}{2kT}(I_{\text{det}} + I_{\text{FET}})\right]\right\}^{1/2}\quad(11)$$

Equation 11 demonstrates

(a) advantages in cryogenic operation of the detector (lower I_{det}), the FET and the feedback resistor

(b) importance of the FET figure of merit g_m/C

(c) need for high-megohm resistors for high resolution preamplifiers

(d) deterioration of resolution for large area detectors (their capacitance is proportional to the area)

(e) opposite dependence on τ of the series and parallel noise.

Assuming the following values for a silicon X-ray spectrometer,

$C_{\text{det}} = 1$ pF, $C_{\text{FET}} = 3.5$ pF, $C_f = 0.5$ pF, $R_f = 5.10^{10}\Omega$, $g_m = 10$ mA/V, $I_{\text{det}} = 0.5$ pA, $I_{\text{FET}} = 0.1$ pA, $T = 100$ K, $\tau = 5$ μs, and using Eq 11 ($\epsilon = 3.8$ eV, $q = 1.6\ 10^{-19}$ Coul) we get $\text{FWHM}_{\text{electronic}} = 96$ eV (Si).

The contribution of the FET (series noise) is 69.5 eV, the feedback resistor 40 eV, leakage current of the detector 48 eV, and the leakage current of the FET 21.5 eV. The various contributions sum up as the square root of the sum of the squares (Eq 11), to give the total electronic resolution of 96 eV.

Using the same example for the opto-electronic feedback (Eq 8 into 11) and taking the quiescent condition ($I_s = 0$) we find that the total noise is 97.5 eV. The electronic noise of the same system when detecting iron X-rays (6.4 keV) at a rate of 10,000 cps ($I_s = 2.7$ pA) is 148 eV.

Equation 11 points out the existence of an optimum time constant (τ_{opt}), at which the resolution is minimum and which is given by

$$\tau_{opt} = \left\{ \frac{0.6(C_{det} + C_{FET} + C_f)^2}{g_m \left[\dfrac{1}{R_f} + \dfrac{q}{2kT}(I_{det} + I_{FET}) \right]} \right\}^{1/2} \quad \ldots\ldots\ldots\ldots (12)$$

Using the previous example $\tau_{opt} = 5.25$ μs. For $\tau > \tau_{opt}$ the parallel noise dominates (resistor and leakage currents) and for $\tau < \tau_{opt}$ the series noise is dominant (FET). A decrease in the parallel noise (for example, pulsed feedback) increases τ_{opt} and consequently improves the resolution. However, the lower noise is obtained at the expense of count-rate performance as the wider output pulses (see section on the amplifier) mean higher probability for pulse pileup. Therefore, for best resolution the system should be operated near τ_{opt}, but for high count-rate applications (10^4 cps) it must be operated with $\tau < \tau_{opt}$. At these time constants the resolution slope varies according to $\tau^{-3/2}$ and depends mainly on the characteristics of the FET. Microphonic noise was not included in Eq 7 and therefore does not appear in Eq 12. Actually, however, microphonics are a parallel noise source, and their presence decreases τ_{opt} and increases the minimum value of the noise.

A typical plot of resolution versus τ for RC and Gaussian filters is shown in Fig. 12. It is clear from that figure that the Gaussian shaper is preferred for low-noise performance, and it is also superior to the simple CR-RC filter as far as high count-rate capability is concerned. The optimum noise width for the tested system was 114 eV at $\tau = 8.5$ μs, (peaking time of 18.5 μs). The noise width deteriorates to only 187 eV at $\tau = 1$ μs, which permitted us to obtain total resolution of 270 eV for iron X-rays at 100,000 cps with no pileup rejection.

Figure 13 shows the resolution versus τ for three different systems in order to demonstrate the relative importance of series and parallel noise (see Eq 7). Curve 1 represents a system with low parallel and series noise (12 mm² × 3 mm detector). The optimum τ is 8.5 μs. Curve 2 is for a system with larger series noise (larger detector: 75 mm² × 3 mm) but still low parallel noise. The optimum τ for that curve is higher than 10 μs as expected, and its slope at lower τ's is higher than for Curve 1. The optimum noise for Curve 2 should be higher than for Curve 1. Curve 3 represents a system with high parallel noise (high I_{FET}) but small series noise.

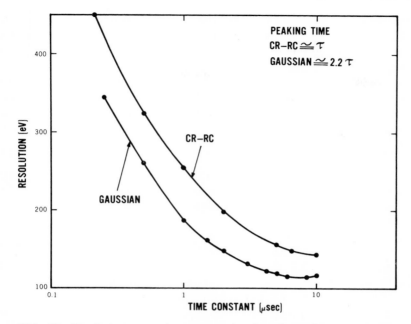

FIG. 12—*Resolution versus time constant for Gaussian and CR-RC shapers.*

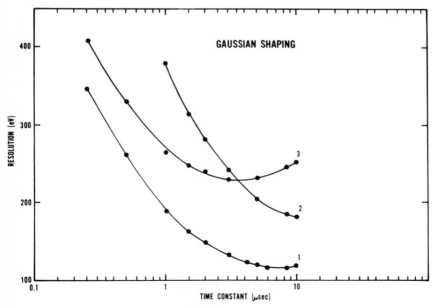

1 • low-series and parallel noise.
2 • low-parallel noise, higher-series noise (larger detector).
3 • low-series noise, higher-parallel noise (FET with high-
    leakage current).

FIG. 13—*Resolution versus time constant for three different spectrometers.*

The optimum τ is around 3 μs and the optimum noise is quite high, 230 eV. However, because of the small series noise the slope at low τ's is low (even lower than for Curve 1) and therefore for very high count-rates (10^5 cps) System 3 may be preferred over System 2.

The total resolution for a certain X-ray line of energy E is obtained by summing the electronic noise, the contribution of incomplete charge collection in the detector and the statistical fluctuation due to the detection process.

$$FWHM^2_{total} = FWHM^2_{electronic} + FWHM^2_{charge\ collection} + 5.52\ \epsilon FE \quad ..(13)$$

where F is Fano factor (typical measured values at low energy X-rays are about 0.13 for both silicon and germanium).

Using our previous example, and assuming perfect charge collection, the total resolution for iron X-rays is

(a) resistive feedback: 164 eV, and

(b) opto-electronic feedback: 198eV.

Experimental results confirm the theoretical calculations. Figure 14 shows a spectrum of manganese X-rays taken with ORTEC's 12 mm² × 3 mm silicon detector and 117 preamplifier with modified resistive feedback. The resolution for the $K\alpha$-line was 152 eV with peak-to-valley ratio approximately 400. The full width at tenth maximum (FWTM) resolution is 282 eV confirming good charge collection of the detector (FWTM = 1.855 FWHM). The electronic noise was 82 eV with Gaussian shaping and a peaking time of 22 μs, while the pulser resolution was 83 eV. For larger area detectors (80 mm²) the resolution for the manganese $K\alpha$-line was 176 eV for the same type of shaping. Comparing the resolutions for the 12 and 80 mm² detectors it is clear that the noise sensitivity to input capacitance is now low enough to reconsider the efficiency-resolution compromise in choosing a detector size for a particular application.

The resolution of germanium X-ray spectrometers has previously been poorer than that of silicon spectrometers, although some predictions and measurements show that the ϵF product for germanium is smaller than that for silicon. Figure 15 shows what we believe to be the closest a germanium system has come to outperforming the silicon spectrometer. The resolution for the $K\alpha$ manganese X-rays was 168 eV as recorded with ORTEC's 30 mm² × 5 mm Ge(Li) detector and 117-A preamplifier. The peak-to-valley ratio here is only 130, and the peak-to-tail ratio is much poorer than for the silicon detector, which is not surprising considering the more serious dead layer problem with germanium. The electronic noise contribution was 128 eV for 7.5 μs peaking time with a Gaussian shaper. It is clear that the ϵF product for the geraminum detector is the smaller one and therefore the real potential of the germanium spectrometer is for X-rays above 10-15 keV. The smaller ϵ for germanium (see Eq 1) suggests the possibility of lower noise widths for germanium spectrometers

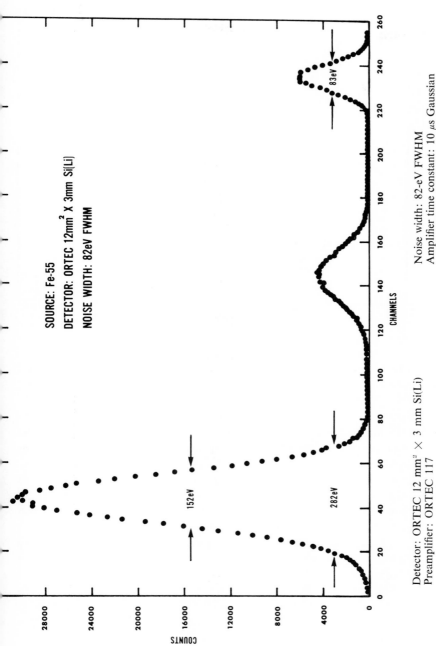

SOURCE: Fe-55
DETECTOR: ORTEC 12mm² X 3mm Si(Li)
NOISE WIDTH: 82eV FWHM

Detector: ORTEC 12 mm² × 3 mm Si(Li)   Noise width: 82-eV FWHM
Preamplifier: ORTEC 117   Amplifier time constant: 10 μs Gaussian

FIG. 14—*Spectrum of manganese X-rays taken with a silicon detector.*

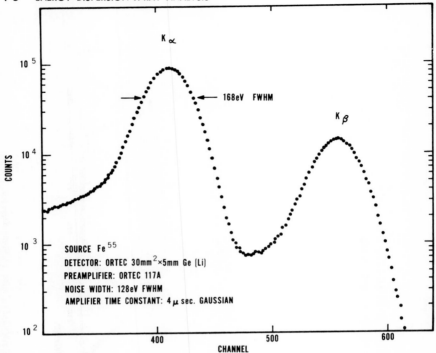

Detector: ORTEC 30 mm² × 5 mm Ge(Li)
Preamplifier: ORTEC 117A
Noise width: 128-eV FWHM
Amplifier time constant: 4 μs Gaussian

FIG. 15—*Spectrum of manganese X-rays taken with a germanium detector.*

than for silicon spectrometers, if the noise sources, and especially the leakage current of the detector, could be made comparable in both cases (see Eq 11).

The very low noise of the silicon spectrometer (82 eV) enhances its use in analysis of low Z materials. Figure 16 shows a spectrum of aluminum (1.49 keV) and silicon (1.74 keV) X-rays produced by fluorescence with an X-ray tube. The resolution of these lines is 109 and 111 eV, respectively, with the two being almost completely separated (peak-to-valley ration 13 to 1). The FWTM resolution is 210 eV pointing out the good charge collection in the detector at these low energies. The noise limit of the spectrometer is well below 500 eV, and thus the detection of oxygen X-rays ($Z = 8$, $E_\alpha = 523$ eV) that will penetrate the entrance window is clearly possible. Figure 16 points out the enhanced application possibilities of such high resolution systems for analysis of low Z elements, for example, on electron microscopes in semiconductor device research.

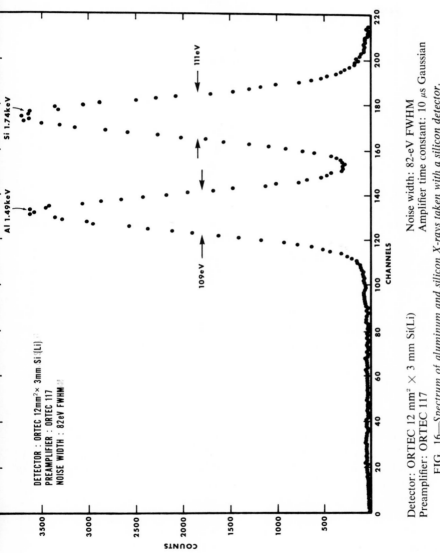

Detector: ORTEC 12 mm² × 3 mm Si(Li)   Noise width: 82-eV FWHM
Preamplifier: ORTEC 117   Amplifier time constant: 10 μs Gaussian

FIG. 16—*Spectrum of aluminum and silicon X-rays taken with a silicon detector.*

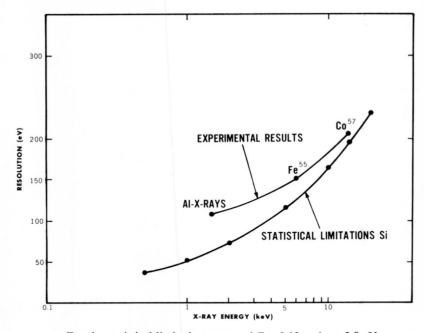

For the statistical limitations assumed $F = 0.13$ and $\epsilon = 3.8$ eV

FIG. 17—*Theoretical limitations and experimental results for a silicon X-ray spectrometer.*

The basic resolution limitation of X-ray spectrometers is the statistical fluctuation in charge production (see Eq 13). Figure 17 displays the limit values for the resolution of a silicon detector (assuming $F = 0.13$ and $\epsilon = 3.8$ eV), and recent experimental results. Future improvements in electronic resolution certainly will be welcomed for analysis of spectra below 5 keV.

Conclusions

Low-noise cryogenic preamplifiers have made tremendous progress during the last four years, thereby opening the field of X-ray spectrometry to semiconductor detectors. However, further improvements in resolution of these preamplifiers are definitely expected to enhance the use of the semiconductor energy analyzers to analysis of low Z materials and their operation at higher count-rates and to improve the accuracy in quantitative measurements. The main improvement is expected from use of new FET's, perhaps from materials other than silicon, although the theoretical capabilities of silicon junction FET's have not been reached yet. Also, new feedback methods, like the pulsed feedback approach, will reduce the parallel noise and therefore improve the resolution at longer time constants (low count rates).

The improving performance of the preamplifier will no doubt bring improvements in detector and cryostat design for low energy X-ray analysis. They may manifest themselves in better charge collection properties, lower leakage currents or smaller Fano factors for the detector, higher transmission windows, and more reliable operation for the cryostat.

The progress made in the design of low-noise cryogenic preamplifiers for X-ray detectors should have a definite impact on other areas where low-noise amplification is desirable.

Acknowledgments

The author wishes to acknowledge the helpful review of this manuscript by F. J. Walter and the technical help of G. R. Dyer. Invaluable help from ORTEC's drafting department, Lloyd Austin, and M. Freels in the preparation of the manuscript and its illustrations is greatly appreciated. Thanks are also due to the Technical Information Division of Lawrence Radiation Laboratory, Berkeley, California, for supplying some of the figures.

References

[1] Elad, Emanuel, *Nuclear Instruments and Methods,* Vol. 37, 1965, p. 327.
[2] Elad, Emanuel and Nakamura, Michiyuki, *Nuclear Instruments and Methods,* Vol. 41, 1966, p. 161.
[3] Cameron, J. F. and Rhodes, J. R., "Filters for Energy Selection in Radioisotope X-ray Techniques" in Clark, G. L. ed., *Encyclopedia of X-Rays and Gamma Rays,* Reinhold, London and New York, 1963, p. 387.
[4] Dothrie, J. J. and Gale, B. K., *Spectrochim,* Acta 20, 1964, p. 1735.
[5] Elad, Emanuel and Nakamura, Michiyuki, *Transactions on Nuclear Science,* Institute of Electrical and Electronic Engineers, Vol. 14, No. 1, 1967, p. 523.
[6] Bowman, H. R. et al, *Science,* Vol. 151, 1966, p. 562.
[7] Elad, Emanuel and Nakamura, Michiyuki, *Nuclear Instruments and Methods,* Vol. 42, 1966, p. 315.
[8] Elad, Emanuel, Ph.D. thesis, University of California, Berkeley, June 1968.
[9] Grove, A. S., *Physics and Technology of Semiconductor Devices,* Wiley, New York, 1967, p. 243.
[10] Sevin, L. J., Jr., *Field Effect Transistors,* McGraw-Hill, New York, 1965.
[11] Elad, Emanuel and Nakamura, Michiyuki, *Transactions on Nuclear Science,* Institute of Electrical and Electronic Engineers, Vol. 15, No. 1, 1968, p. 283.
[12] Radeka, Veljko, Brookhaven National Laboratory, Report BNL-12748, 1968.
[13] Goulding, F. S., Walton, J. T., and Pehl, R. H., Lawrence Radiation Laboratory, University of California, Berkeley, Report UCRL-19377, Oct. 1969.
[14] Kandiah, K. L. and Stirling, A. D., *Semiconductor Nuclear Particle Detectors and Circuits,* National Academy of Sciences, Publication 1593, Washington, D. C. 1969, p. 495.
[15] Radeka, Veljko, Brookhaven National Laboratory, Report BNL-14492, 1970.
[16] Elad, Emanuel, *Proceeding Ispra Nuclear Electronics Symposium,* 1969, p. 21.
[17] Miner, C. E., *Nuclear Instruments and Methods,* Vol. 55, 1967, p. 125.
[18] Fairstein, Edward and Hahn, John, *Nucleonics,* Vol. 23, No. 11, 1965, p. 50.
[19] Fairstein, Edward and Hahn, John, *Nucleonics,* Vol. 24, No. 1, 1966, p. 54.
[20] *Semiconductor Nuclear Particle Detectors and Circuits,* National Academy of Sciences Publication 1593, Washington, D. C., 1969, Chapter VIII-Filters, p. 509.

F. J. Walter[1]

Characterization of Semiconductor X-ray Energy Spectrometers

REFERENCE: Walter, F. J., **"Characterization of Semiconductor X-ray Energy Spectrometers,"** *Energy Dispersion X-ray Analysis: X-ray and Electron Probe Analysis, ASTM STP 485,* American Society for Testing and Materials, 1971, pp. 82–112.

ABSTRACT: The important performance parameters of a semiconductor X-ray energy spectrometer and the standardized techniques for measuring and documenting most of the related specifications are described.

The strong coupling between performance parameters such as energy resolution, detector size (efficiency), count-rate capability, and general spectral integrity (tailing, etc.) is detailed. In particular, the conflict between the need to improve energy resolution by increasing the amplifier shaping time constants (that is, minimizing field-effect transistor noise) and the need to improve count-rate capability by decreasing amplifier pulse shaping time constants are examined thoroughly.

Since, as is shown, most performance parameters can be improved at the expense of other characteristics, the importance of providing enough test data to determine the trade-off cost involved in optimizing a particular variable is emphasized.

KEY WORDS: semiconductors (materials), semiconductor devices, X-ray spectrometers, X-ray analysis, preamplifiers, amplifiers, resolution, performance

Foreword

Much of the information in the sections of this presentation which deal with standardized methods for measuring and specifying performance parameters has been abstracted from preliminary working drafts of a set of Standard Test Procedures which are being prepared by the Nuclear Instruments and Detectors Committee (NIDCOM) of the Institute of Electrical and Electronic Engineers (IEEE) Nuclear Science Group in cooperation with the American Society for Testing and Materials (ASTM) Subcommittee XV of Committee E-4 on Metallography. Comments or suggestions on Standard Test Procedures should be directed to G. L. Miller, chairman, NIDCOM, J. C. Russ, secretary, ASTM Subcommittee XV, or F. J. Walter, project leader for IEEE Standard Test Procedures, Semiconductor X-ray Energy Spectrometers.

[1] Manager, Semiconductor Research and Development, ORTEC, Inc., Oak Ridge, Tenn. 37830.

For the reader who is interested in more information about standard test procedures [1],[2] we suggest starting with IEEE Standard No. 300, USAS N 42.1, 1969, and IEEE 301, USAS N 42.2, 1969, since the symbols, glossary, definitions, and general approach used herein are based on these documents. IEEE 300 and 301 were prepared originally to deal with the problems of Standardized Test Procedures for Semiconductor Detectors and Amplifiers Used for Nuclear Physics Research.

A semiconductor X-ray energy spectrometer is an instrument which utilizes the proportionality between the energy of an X-ray and the number of free electron-hole pairs created in a semiconductor by that X-ray and the subsequent motion of these carriers as a means for measuring the energy of the X-ray. In this presentation, the spectrometer is assumed to consist of the semiconductor detector, preamplifier, means for cooling same, encapsulation, required vacuum and mechanical housings, amplifier, power supplies, and other required analog electronics. For the purposes of this discussion, analog-to-digital data conversion and subsequent data handling equipment are not considered to be a part of the spectrometer. In addition, specimen stages, the X-ray emitter, and means for exciting same (for example, radioactive sources, electron beams, etc.) are considered as ancillary equipment and not part of the spectrometer.

Introduction

In its simplest analogy, the semiconductor detector is analogous to a gas counter in which the gas has been replaced by a solid semiconductor. Figure 1 illustrates how this type of detector operates. The detector consists of two conducting electrodes with the region between filled with a semiconductor crystal. A voltage difference is placed across the electrodes thereby producing an electric field in the semiconductor. When an X-ray enters the semiconductor and is totally absorbed, it loses its energy by producing free charge carriers in the semiconductor. These carriers, the number of which is proportional to the energy of the X-ray, move under the influence of the electric field until they are collected at the electrodes or trapped internally in the crystal. The resulting current represents the basic signal information, and the integrated current is proportional to the energy lost by the X-ray. The signal is amplified and shaped to produce a pulse whose amplitude is proportional to the energy of the incident X-ray. These pulses are routed to a multichannel pulse height analyzer (MCA) where they are sorted and stored according to amplitude. The output of the MCA constitutes the complete X-ray spectrum.

The semiconductor materials (that is, silicon and germanium) which are used currently for X-ray spectrometers do not have sufficiently high resistivity to withstand large electric fields without excessive leakage currents.

[2] The italic numbers in brackets refer to the list of references appended to this paper.

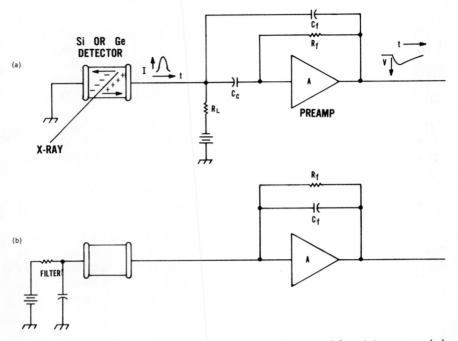

FIG. 1—*Schematic of semiconductor detector and preamplifier.* (a) *a-c coupled configuration,* (b) *d-c coupled configuration.*

This makes it necessary to use special techniques to limit the current flow through the semiconductor device. This is accomplished by utilizing the space charge region of a reverse biased diode junction as the detector sensitive volume. Silicon and germanium are the only semiconductor materials which are presently available in sufficiently perfect and pure single crystals to be useful for this type of device. Although both of these materials are available in very pure form, that is, with electrically active impurities in the parts per billion range, the electrically active impurity level is usually still too high to allow one to make a detector with a large volume in the depletion region of the diode junction. This problem is overcome frequently by using the lithium drift process which is a technique of drifting in mobile lithium ions to compensate for the electrically active impurities. If this technique is used, the semiconductor device is subject to some inherent stability problems: The compensation by lithium is not stable at room temperature. In germanium the compensation by lithium can be classed as highly unstable at room temperature, and, therefore, lithium compensated germanium devices must be stored at reduced temperatures. In silicon the situation can be termed quasi-stable with the stability depending on the presence of other impurities in the crystal, the details of the device fabrication procedure, and other ambient conditions. The semiconductor detector and first stage of the preamplifier are operated usually at low temperature to reduce the thermal noise [2]. Since there is no internal gain in the type of semiconductor detector considered here, the problem of noise in the preamplifier circuits is significant, particularly at low X-ray energies.

The following sections describe the parameters which affect resolution, efficiency, count-rate capability, linearity, spectral fidelity, etc. and the trade-offs between them.

Energy Resolution and Spectral Distortion

Energy resolution is one of the most important characteristics of an X-ray energy spectrometer since it sets the limit on the ability to resolve the peaks from two adjacent elements. The lighter the element, the better the resolution required. This is illustrated in Fig. 2 which shows the spacing between adjacent elements or elemental lines as a function of atomic number. However, in comparing instruments, it is necessary also to examine the other specifications since improved energy resolution can be obtained usually at the expense of other important parameters such as count-rate capability, gain stability, and detector size (efficiency). The limit resulting from the statistical uncertainty involved in the process of converting the incident X-ray energy into free charge carriers is the most funda-

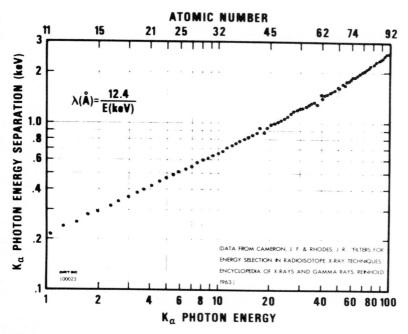

FIG. 2—*Spectral line separation as a function of X-ray energy and atomic number.*

mental limit on energy resolution. Preamplifier noise and noise in the detector also limit the resolution. Charge loss from trapping and recombination also can have a significant effect on the spectral quality. The peak width broadening from charge loss increases with increasing energy since it is usually a percentage charge collection effect [3]. "Partially dead" regions or layers on the detector also can produce spectral distortion in the form of low energy tails, nonlinearities, peak shape distortion, and loss of efficiency at low energies [4].

Electronic stability is also important, and this includes not only the stability of the preamplifier electronics associated directly with the detector, but also the stability and quality of other ancillary instruments such as the multichannel analyzer. Obviously, extraneous sources of noise and disturbance also can cause spectral deterioration.

Effects of high count rate and pulse pileup are also important. Every time an event produces a pulse in the amplifying equipment, the d-c levels throughout the system are perturbed and take some time to return to their original level. If another event should occur within this time interval, its effective output pulse height may be distorted, thereby contributing to the spectral distortion.

The resolution of an energy analyzing detector is described customarily by the full width at half maximum (FWHM) and full width at one tenth maximum (FWTM) of a monoenergetic X-ray spectral line. FWHM is defined [1] as the full width of a distribution measured at half the maximum ordinate: for a normal distribution, it is equal to $2(2 \ln 2)^{1/2}$ times the standard deviation σ. Similarly the FWTM is the full width at one tenth the maximum ordinate.

Statistical Limit of Resolution

If Poisson (true random) statistics applied to the process of the X-ray energy loss mechanism, then it would be possible to calculate the statistical broadening of a monoenergetic X-ray peak from FWHM $= 2.35 \sqrt{\epsilon E}$, where E is the energy of the incident radiation and ϵ is the average energy required to create a free electron hole pair, that is, the number of free charge carriers created is the incident energy over epsilon ($N = E/\epsilon$). (The value of ϵ at 77 K presently is assumed to be 3.8 eV/pair for silicon and 3.0 eV/pair for germanium) [5,6]. In fact, this process does not follow Poisson statistics because the multiple events that occur in the energy loss process are correlated, and therefore the form of the statistical formula must be modified. This is done usually by inserting a correction factor, called the Fano factor (F), resulting in the expression FWHM $= 2.35 \sqrt{\epsilon F E}$. This corrects for the fact that the real broadening is not as great as would be predicted by Poisson statistics (that is, the Fano factor is less than unity and for silicon and germanium is presently thought to be between 0.05 and 0.15) [6,7].

Preamp Noise

Preamp noise arises from a number of sources.[3] One source is the thermal noise in the first field-effect transistor (FET). This is minimized usually by cooling the FET along with the detector and by using specially selected FET's. The effective noise generated by the field-effect transistor increases with the total capacitance at the FET gate and therefore increases with detector area. Thus, there is an important trade-off between detector size and preamp noise. This noise is also sensitive to parasitic stray capacitance; consequently, the design layout of the detector and preamp package is quite critical. The resistors R_f (and R_L) shown in the circuit of Fig. 1a are also important sources of noise. In order to minimize noise resulting from R_L, C_c and R_L frequently are eliminated. However, in order to make this work it is necessary to d-c couple the detector directly into the gate of the first FET. Thus, all of the average current caused by the radiation must flow through the feedback resistor R_f, since changing the current flow through the high impedance gate of the field effect transistor would change its operating point and therefore the overall gain of the system. The larger the feedback resistor R_f the smaller the resistor noise but the poorer the count-rate capability. This presents an important trade-off consideration between noise broadening of the resolution at low count rates and performance at high count rates. In this respect, it should be pointed out that it is not enough to specify just the simple count rate capability because the problem of d-c shift at the FET gate is proportional to the total radiation induced current flow. Thus, it is the count rate times the average radiation energy which is important [4], for example, 100 keV X-rays present 20 times the count rate problem of 5 keV X-rays. Other feedback mechanisms such as optical feedback [8] and intermittent d-c (pulsed) feedback [9] are used sometimes in place of R_f. However, these techniques also can involve resolution versus count rate compromises.[3]

Figure 3 shows the combined effects of preamp noise and statistics for several assumptions about the Fano factor and preamp noise. Since it is assumed that these factors are uncorrelated, these effects are added in quadrature. The preamp contribution is flat and independent of energy, while the statistical broadening varies as the square root of the energy.

Detector Noise

Leakage current in the detector element also can contribute to system noise and resolution broadening. In integrated X-ray spectrometer systems, it is usually not possible to distinguish unambiguously between series noise sources (for example, the FET) and parallel noise sources such as detector leakage current, feedback resistor, and microphonics. However, the

[3] See p. 57.

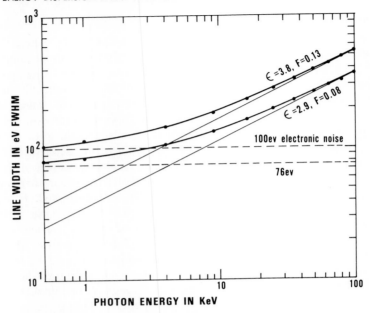

FIG. 3—*Illustration of relative contributions of electronic noise and statistical broadening to total line width.*

parallel noise usually predominates at lower frequencies, so some indication of the presence of detector current noise and resistive feedback noise can be obtained by examining the noise as a function of amplifier passband. At the present state of the art, detector leakage current can be a negligible factor for detectors which are cooled to liquid nitrogen temperature (78 K). However, since the detector noise increases more rapidly with increasing temperature than does the FET noise, detector noise can be expected to become a major contribution at temperatures above about 200 K.

Charge Loss

Loss of free charge carriers because of trapping throughout the semiconductor crystal can be a significant source of spectral distortion. The peak width broadening from this effect increases with increasing energy of the X-ray and, therefore, will be more pronounced at higher energies. This problem, which is often a function of detector bias voltage, frequently results in low-energy tails on monoenergetic spectral lines, and, therefore, is often quantified more readily by changes in the FWTM than in the FWHM of the spectral peak. The peak-to-valley ratios for $K\alpha$ and $K\beta$ X-ray lines are also quite sensitive to this effect. The FWTM performance of the system is more important than it might appear on superficial examination. For example, if it is desired to determine a 10 percent X-ray fluorescence contribution from one element in the presence of a 90 percent contribution from the next higher element, the FWTM is more relevant than the FWHM.

Windows and dead layers which are not an intimate part of the detector element cause a loss of efficiency at low X-ray energies. Charge loss in "partially dead" [4] layers in the detector adjacent to the entrance window can result in both a loss of full energy peak efficiency and in undesirable spectral distortion in the form of low energy tails, ghost peaks, and non-linearities.

Electronic Stability and Extraneous Noise

The gain stability of the system, as well as the presence of excess noise from spurious electrical effects or microphonics, are important aspects of the performance of the system. The FWTM is sometimes more sensitive to these problems than the FWHM.

Procedures for measuring electronic gain stability as a function of ambient temperature and line voltage are described in Ref 1.

Count-Rate Effects

Pulse Shape Considerations and Distortion by Pulse Pileup

In order to optimize the signal-to-noise ratio as well as to provide reasonable count-rate capability, it is necessary to restrict both the upper and lower limits of the amplifier passband.[3] At any given count rate, the width (in time) of the shaped pulse will determine the probability that the preceding or succeeding pulse will be distorted by pulse-on-pulse or pulse-on-tail pileup. Choice of the type of pulse shaping and time constants associated with the pulse therefore represent an important trade-off between optimum energy resolution at low count rate and the performance of the system at high count rates. Preamplifier noise frequently can be minimized by using relatively long pulse shaping time constants. The resulting optimum shaping time constant for minimum noise, therefore, can be inconsistent with the count rate requirements of the user. Consequently, energy resolution specifications should be always accompanied by either the applicable count rate, detailed information on the pulse shape, or a detailed description of the change in FWHM and FWTM as a function of count rate at a specified energy. The latter description of performance is particularly useful in cases where the instrument is to be used in quantitative analysis where changes in resolution with changing count rate would seriously complicate analysis of the resulting data.

The usual shaping time constant number found on the front panel of the amplifier is by itself of little use in predicting count-rate capability. To first order, it is the width of pulse near the baseline which determines the count-rate pileup capability of the system. For a given nominal time constant, this width can vary by a factor of two or more depending on the type of pulse shaping and the convention used for assigning a time constant value.

The requirement that the characteristic width of the pulse be small compared to the mean interval between pulses is a necessary but not sufficient requirement for good energy resolution. It is also necessary that the pulse return to the baseline (average d-c level) in a time interval small compared to the mean interval between pulses. Most preamplifier-amplifier systems utilize a design principle referred to as pole-zero (PZ) cancellation to minimize undesirable undershoots [10]. Failure to achieve accurate PZ cancellation can degrade the FWHM and FWTM and also can result in low-energy tails or high-energy tails on the peak depending on whether an undershoot or overshoot exists.

Baseline (d-c level) Shifts, D-C Restoration, Count-Rate Induced Gain Shifts, and Nonlinearities

A count-rate induced d-c level shift anywhere within the electronic amplifying chain will introduce gain shifts and nonlinearities whenever the d-c level exceeds the linear dynamic range of that portion of the signal processing chain. Also, once the linear dynamic range is exceeded, statistically induced fluctuations in the d-c level result in peak broadening and other related spectral distortions. In addition, a short-term d-c level shift at the output of the amplifier system will produce an equivalent shift in the magnitude of the output voltage pulse presented to the analog-to-digital converter (ADC) or single channel analyzer and thereby a broadening of the resolution.

Customarily, these problems are compensated for by using heavy d-c feedback and baseline restoration (BLR) at the output of the amplifying chain [11]. It was pointed out previously that the feedback in preamplifiers can be accompanied by increased noise.[3] D-C restoration (that is, BLR) also can be accompanied by an excess noise contribution which depends on the restoration mode and restoration rate.

When specifying resolution, stability, gain drift, linearity, etc., as a function of count rate, it is essential that the following supporting information be available:

1. The energy for which the performance is specified (count-rate problems are related frequently to the product of rate times energy).

2. Details of the pulse shape and restoration mode; a simple statement of "amplifier time constant," "peaking time," etc., is of very little value unless accompanied with enough information to allow one to determine the width of the pulse at the baseline.

3. Whether relative specifications as a function of count rate are all obtained at the same operating conditions (for example, same pulse shape, feedback configuration, restoration mode, etc.).

Consideration [3] is particularly relevant to applications where it is not practical to reoptimize the spectrometer system every time the product of the count rate and the energy changes. Figure 4 illustrates what some typical resolution versus count-rate curves might be expected to look like. A nonzero slope in the portion of the curves labeled AB can be caused by uncanceled long time constant poles, feedback noise which increases with increased count rate, and inadequate d-c restoration. The shape of section BC is affected by short time constant uncanceled poles, inadequate baseline restoration, tails or undershoots on the pulse, and by the existence of a section of the amplifying chain where the linear dynamic range has been exceeded. Section CD is dominated usually by pulse-on-pulse pileup. The first two sections are determined mostly by the design and therefore, with the exception of choice of BLR mode, are usually beyond the control of the user. Characteristics of section CD, however, are related directly to the width of the shaped pulse (τ) and, therefore, can be changed easily on systems which are provided with adjustable amplifier time constants. Since count-rate distorted spectral peaks do not retain their Gaussian shape, a curve such as Fig. 4 tells only part of the story. Consequently, it is usually desirable to provide a plot of FWTM as well as FWHM. Figures 5 and 6 illustrate some typical count-rate induced spectral distortions. The data in Fig. 5 were obtained at moderate count rates with and without baseline restoration. Figure 6 compares the results of good and inadequate PZ cancellation. In the latter case, a PZ cancellation network in the amplifier was intentionally slightly detuned in order to simulate the effect of a poorly compensated pole.

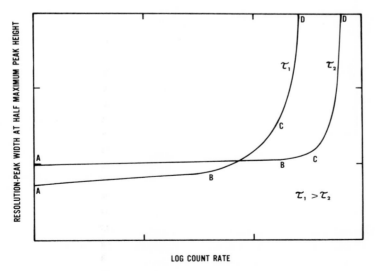

FIG. 4—*Energy resolution as a function of count rate.*

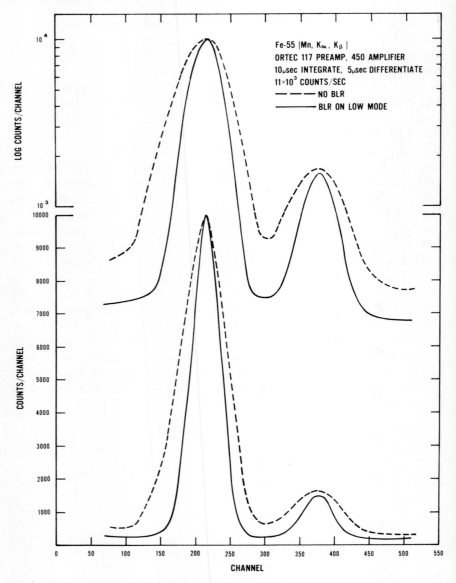

FIG. 5—*Spectral distortion caused by lack of adequate baseline restoration.*

Overload Recovery

Overload pulses produced by high-energy radiation events which exceed the dynamic range of the amplifying equipment can produce undesirable effects such as baseline overshoot or undershoot and system paralysis (dead-

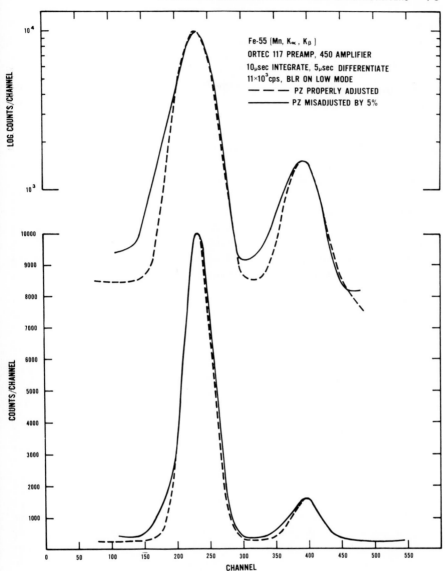

Fe-55 (Mn, K$_\alpha$, K$_\beta$)
ORTEC 117 PREAMP, 450 AMPLIFIER
10μsec INTEGRATE, 5μsec DIFFERENTIATE
11×10^{3}cps, BLR ON LOW MODE
– – – PZ PROPERLY ADJUSTED
——— PZ MISADJUSTED BY 5%

FIG. 6—*Spectral distortion resulting from poor cancellation of a pole in the amplifier.*

time). Since designing for good overload recovery performance will compromise other system characteristics it should not be overemphasized except in those applications where high-energy background or pulse-on-pulse pileup are expected to create overload signals.

Measurement of Energy Resolution and Count Rate Distortion of Resolution

X-ray Reference Sources

In order to standardize test procedures which are independent of various sources of X-rays which might be used eventually with the instrument, radioactive sources which are readily available to all manufacturers and users are recommended for specification and intercomparison purposes. Whenever possible, isotopes which are free from problems caused by interfering lines should be selected. This is particularly important in resolution versus count-rate tests where the presence of several strong lines will complicate interpretation of the results or where overload from a high-energy line might distort the results. Table 1 lists a number of useful radioactive sources which cover the energy range of interest for both silicon and germanium detectors when used with characteristic elemental X-rays. It is recognized that special applications or requirements may necessitate the use of additional X-ray energies to adequately characterize an instrument.

TABLE 1—*Useful radiactive sources for testing X-ray energy spectrometers.*

Source	Type	Energy (keV)	Half-Life	Comments
Fe-55............	Mn $K\alpha$	5.9	2.6 years	preferred standard
Co-57............	Fe $K\alpha$	6.4	270 days	not recommended at 6.4
	gamma	14.4		or for count-rate tests
	gamma	122		
Am-241..........	gamma	26.4	458 years	assorted lower energy
	gamma	59.6		gammas and Np X-rays; not recommended for low-energy or for count-rate tests

If only one energy is used to characterize a silicon X-ray spectrometer, Fe-55 (manganese $K\alpha$) is the preferable source since its energy is high enough to make its spectrum sensitive to charge trapping problems in the detector, but sufficiently low so that the resolution is not entirely dominated by statistics. The minimum amount of information required to characterize the resolution performance of a detector system is one X-ray line (for example, manganese $K\alpha$) and the electronic noise contribution (theoretical width at zero energy) to the width. In principle these two pieces of information allow one to predict an upper limit for the resolution at intermediate energies by adding the two resolution factors in quadrature (see Eqs 6, 7, 8). However, this procedure ignores the fact that the resolution at low energies may be broadened by poor charge collection in partially dead layers at the detector entrance window [4]. Since low-energy X-rays are absorbed preferentially near the entrance window, they are particularly sensitive to poor charge collection in "partially dead" entrance windows. Consequently, additional reference sources below 6 keV (possibly $L\alpha$ sources) may be required eventually to adequately characterize the resolution of a system.

Iron $K\alpha$ from Co-57 has been widely used in the past as a resolution calibration standard. This source, which has prominent lines at 6.4, 14, and 122 keV, would appear to be very convenient since it spans a broad energy range. However, it has not been recommended here because the prominent line at 14 keV complicates interpretation of count-rate effects. In addition, if the system gains are adjusted to make the 6 or 14-keV lines utilize a significant fraction of the dynamic range of the amplifier, the overload events caused by the 122-keV line will distort the spectrum. This effect is illustrated in Fig. 7 which shows two spectra obtained with the same energy spectrometer at the same count rate. One spectra is iron $K\alpha$ produced by X-ray fluorescence of iron and the other is iron $K\alpha$ from Co-57. Notice that even at these modest count rates, the overload pulses from the 122-keV gamma result in a significant loss in resolution.

Simulating the Charge Pulse of a Detector

A simulated pulse from a pulse generator is useful for measuring the electronic noise contribution to the resolution, electronic linearity, gain stability, and for energy calibrations. In order for all sources of noise, drift, etc., to be sampled by the test pulse, the simulated charge pulse must be injected into the *input* of the preamplifier. This charge pulse can be simulated by applying a voltage step to a small capacitor connected in parallel with the preamplifier input. The most common type of pulse generator uses a highly stable voltage reference supply to charge up a large capacitor. The required voltage step is then produced by using a mercury-wetted-contact relay to connect the storage capacitor to the small coupling capacitor. This type of pulse generator frequently is referred to as a tail pulse generator since the voltage step decays away with the RC time constant of the generator storage capacitor and the terminating resistor [1].

Pulse generators which produce a square pulse rather than a long tail pulse also are used occasionally for this purpose.

The exponential decay corresponding to the long tail on the pulse represents an uncanceled pole which will show up as an undershoot at the amplifier output. Pole zero cancellation techniques [1] are not appropriate to this problem since they would result in excess noise at the preamp input. For a square pulse generator, droop in the pulse will constitute a long time constant pole, and, in addition, the negative going portions of the pulse will produce a pulse of the wrong polarity in the amplifier and probable resulting baseline distortion. Consequently the use of pulse generators is not recommended for count-rate effect measurements [1]. Unless it has

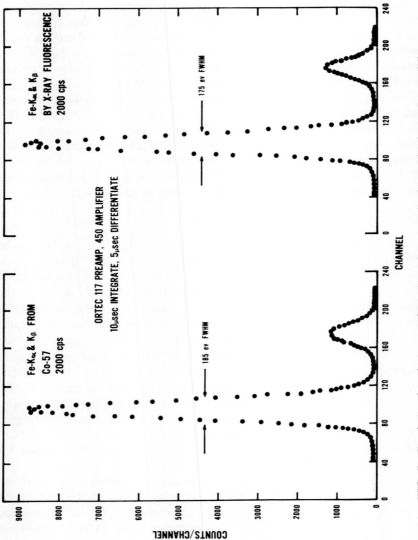

FIG. 7—Resolution broadening resulting from large amplitude overloads in the amplifier.

been established previously that the pulse generator is not distorting the other peaks, it is preferable to turn the pulse generator off during X-ray resolution measurements, particularly if the measurements are done at high count rates or long pulse shaping time constants.

Noise Measurement by Pulse-Height Distribution (preferred method)

The electronics and detector noise contributions to the resolution width can be determined by this method only if the instrument is provided with a pulse generator (pulser) coupling capacitor to the *input* of the preamplifier. The spectrometer is connected to a multichannel analyzer of at least 100 channels. After the appropriate recommended system warm-up time, with all elements of the system operating within their linear range and with the detector at its specified operating voltage, expose the detector to an appropriate X-ray source (for example, 5.9-keV manganese $K\alpha$ from Fe-55) and accumulate an X-ray spectrum. The system gain and pulse amplitude should be such that the FWHM of the peak is at least ten channels (for small MCA's, it may be necessary to use a biased amplifier to achieve this condition). Remove the X-ray source and calibrate the pulser by setting its output amplitude dial to the appropriate X-ray energy (that is, 5.9 keV) and adjusting the calibration control until the pulser peak is in the same peak channel as the X-ray peak. With the source removed, and with the system gain such that the FWHM of the pulser peak is at least ten channels, accumulate two peaks corresponding to pulse generator output energies E_1 and E_2 in channels N_1 and N_2. The system total noise line width (FWHM) is defined [1] as

$$\Delta_E^T = \left(\frac{E_1 - E_2}{N_1 - N_2}\right) \Delta_N^T. \dots\dots\dots\dots\dots\dots\dots(1)$$

where Δ_E^T is the FWHM in units of energy and Δ_N^T is the FWHM in channels. For accurate results, it will be necessary to interpolate between channels. If the results are taken directly from the display (instead of from digital printout), it should be remembered that it is the number of spaces between dots that represents the number of channels.

The FWTM is defined similarly as

$$\delta_E^T = \left(\frac{E_1 - E_2}{N_1 - N_2}\right) \delta_N^T. \dots\dots\dots\dots\dots\dots\dots(2)$$

where δ_N^T is the width of the peak (in channels) at one-tenth maximum height.

If the pulse generator is triggered by line frequency, this technique will not detect the presence of any noise which is correlated with 60 Hz. This problem may be avoided by using a pulse generator with an internal trigger which is nonsynchronous with 60 Hz.

When quoting the total noise line width, complete information on the pulse shaping (including the full pulse width at the baseline) should be stated, for example, 2 μs CR-RC, or semi-Gaussian with four 2 μs CR integrates and one 2 μs RC differentiate. If baseline restoration (BLR) is used, the appropriate characteristics of the BLR should be stated also.

Noise Measurement by Oscilloscope and Root-Mean-Square Voltmeter

An alternative method of measuring noise employs a wide band root-mean-square (rms) voltmeter. The rms noise voltage at the amplifier output is indicated on a voltmeter having a flat band frequency response extending to at least ten times the band center frequency of the amplifier pulse shaping networks. The amplifier incremental gain must be constant down to below the noise level; that is, a biased amplifier section would render this measuring technique useless. A BLR which removes part or all of the negative portion of the noise would also render this technique useless. The system is exposed to X-rays of energy E, and the magnitude of the resulting amplifier output V_a is measured in a calibrated oscilloscope. The X-ray source is removed and the rms noise voltage e_{no} at the amplifier output is read from the voltmeter. The system FWHM noise line width is given by

$$\Delta_E^T = 2.35 \, (a \times e_{no}) \, (E/V_a). \dots\dots\dots\dots\dots\dots (3)$$

where a is 1.13 for a sine wave calibrated average reading voltmeter and a is unity for a true rms voltmeter.

The BLR on most spectrometers eliminates much of the low frequency noise produced by microphonics, 60-Hz pickup, etc. Consequently, in cases where low frequency noise is present, this technique will indicate an apparent noise larger than that actually obtained by the previous method.

Noise Line Width as a Function of Amplifier Time Constants

Information about the relative series (FET) and parallel (detector current, resistor noise, etc.) noise contributions can be obtained by plotting noise line width as a function of amplifier time constants.[3] In addition, such a plot can give important information relative to the count rate versus energy resolution compromise. To perform this measurement, the methods of the preceding paragraphs are employed with an amplifier having adjustable shaping time constants. The results are displayed as a plot of noise line width as a function of pulse shape time constants (with other possible variables such as BLR mode as a parameter).

X-ray Line Width Measurements by Pulse Height Distribution

These tests are carried out using one of the radiation sources described previously. For these cases where the source is a true characteristic X-ray, the location of the centers of the $K\alpha$ and $K\beta$ peaks can be used to obtain the calibration $[(E_1 - E_2)/(N_1 - N_2)]$. For other cases, or if the $K\alpha$ and $K\beta$ are not resolved sufficiently well, the pulser technique described previously may be used to obtain the energy per channel calibration. The results should be expressed as the FWHM and FWTM which are given by

$$\Delta_E^S = \left(\frac{E_1 - E_2}{N_1 - N_2}\right) \Delta_N^S \quad \ldots\ldots\ldots\ldots\ldots\ldots\ldots (4)$$

and

$$\delta_E^S = \left(\frac{E_1 - E_2}{N_1 - N_2}\right) \delta_N^S \quad \ldots\ldots\ldots\ldots\ldots\ldots\ldots (5)$$

where Δ_N^S is the width of the X-ray peak (in channels at half maximum height) and δ_N^S is the analogous width at one-tenth maximum height. Note that if radioactive sources without high-energy lines which perturb the spectrum are used, background subtraction is not required for determining the maximum and one-tenth maximum points. For a perfect Gaussian peak, $\delta_E = 1.8 \times \Delta E$. The FWHM should be at least ten channels wide. In recording information on spectral resolution the following should be noted:

(a) incident X-ray source and energy;

(b) count rate at which measurement was taken;

(c) detector-source geometry;

(d) detailed description of detector, that is, shape, area, depth, silicon or germanium, etc.;

(e) detector operating bias;

(f) system noise (Δ_E^T) line width under the same operating conditions;

(g) if any form of pileup rejection was used, include sufficient details to allow an estimate of the counting loses; and

(h) amplifier pulse shaping parameters (including the full pulse width at the baseline) and BLR parameters.

If information on resolution as a function of count rate is provided, all other variables should be held constant or be plotted as a parameter of a family of resolution versus count rate curves.

Another set of parameters of interest is, Δ_E^O and δ_E^O, those contributions to the detector line width caused by all factors other than electrical noise (that is, statistics and charge collection problems). These quantities are found by subtracting out the noise line width contribution with the relations

$$\Delta_E^O = \sqrt{(\Delta_E^S)^2 - (\Delta_E^T)^2} \quad \ldots\ldots\ldots\ldots\ldots\ldots (6)$$

and

$$\delta_E^O = \sqrt{(\delta_E^S)^2 - (\delta_E^T)^2} \quad \ldots\ldots\ldots\ldots\ldots\ldots (7)$$

It was pointed out previously that the lower limit on Δ_E^O is set by statistical considerations: this limit is given by

$$\Delta_E^O = 2.35\sqrt{FE\epsilon} \quad \ldots\ldots\ldots\ldots\ldots\ldots (8)$$

where:

Δ_E^O = FWHM line width in units of energy,

E = X-ray energy,

ϵ = average energy to form an electron-hole pair, and

F = Fano factor.

Discrepancies between this limit as determined by the smallest experimentally verified Fano factor, and what is observed actually in a particular instrument may be attributed to poor charge collection in the detector. Since no one has been able to unambiguously experimentally separate the problem of charge loss (trapping and recombination) from the statistics of charge formation, state-of-the-art measurements of F must be treated as upper limits on F [4,6,7].

In the preceding, we have suggested repeatedly a lower limit on the number of channels across the peak. Obviously the number of counts per channel (that is, counts per channel in the peak channel) has an equally important bearing on the accuracy and integrity of the results. Clearly, the best accuracy (for a given total number of counts in the spectrum) is provided by a computer fit to the data. There certainly will be a discrepancy between the number of events required for a good subjective visual result and what is required for an equivalent computer fit. Although computer processing of data can effect significant test time efficiency (and therefore cost efficiency) gains, we remain suspicious of overreliance on elaborate and sophisticated processing of meager data; such procedures raise the question of whether the limited amount of data arises from poor count-rate capability or poor gain versus time stability (or both), either of which would represent fundamental weaknesses in the spectrometer.

Table 2 illustrates the amount of data (channels and counts per channel) which is required if one wishes to obtain a desired accuracy in the determination of FWHM or FWTM without resort to computer data fitting schemes. This table is based on the statistical error (95 percent confidence limit) in the height of the maximum channel and the statistical errors in determining the fractional channel position of the half maximum point. Actually, Table 2 is a very conservative estimate, since it completely discounts the ability of the objective eye and mind to function as a very accurate curve fitting computer. However, since it tends to eliminate the objective versus subjective conflict, Table 2 is included as a relevant bench mark on the input data required for reasonably accurate measurement of spectrometer resolution.

Provisions for spectral broadening by gain drifts and count rate related spectral distortions are not included in Eqs 1 through 8. Consequently, before using these formulas, particularly Eqs 6, 7, and 8, it should be established that these effects which do not add in quadrature are negligible contributions to the total width.

TABLE 2—*Requirements for a specified percent accuracy in the measurement of FWHM and FWTM.*

Desired Percent Accuracy in FWHM or FWTM Measurement	Minimum Height of Peak (counts)	Minimum Number of Channels in FWHM or FWTM
1	40 000	40
2	10 000	20
5	1 700	8
10	400	4

In the preceding tests, it is essential that the characteristics of the MCA, such as pulse shape compatibility, nonlinearity, gain stability, and count-rate capability, not substantially distort the results.

Because charge trapping problems can be sensitive functions of energy, it is frequently useful to provide information on the energy resolution at different energies. When this type of information is suppled, it is important that all other variables (for example, count rate, amplifier time constant, etc.) either remain constant or are plotted as parameters.

Spectral distortion from charge trapping problems in the semiconductor also can be sensitive to the amplifier time constants, particularly in deep detectors which are operated at low bias voltages. Therefore, measurement of the X-ray resolution (particularly FWTM) can provide insight into charge trapping problems.

Semiconductor X-ray energy spectrometers can gather data at a very high rate if some compromise in energy resolution is accepted. This is illustrated in Fig. 8, which shows a spectrum which was obtained at a count rate of 10^5 counts per second (cps). The preamp used in this example has a count-rate energy product capability of 2×10^6 keV/s. The same system was capable of 165-eV FWHM resolution at low count rates (Fe-55) by simple switch adjustment of the amplifier time constants and BLR mode. However, the ultimate in resolution at low count rates frequently involves major changes in the components and circuit within the cooled assembly (cryostat). Modification of this part of the system is an expensive and time consuming factory job. Consequently, the user who desires the best of the state of the art may find himself faced with a difficult decision in the resolution versus count-rate capability compromise. Obviously, this decision can be made only in the context of the application for which the instrument is intended.

Peak-to-Tail and Peak-to-Valley Ratios

There are a number of types of spectral distortion which are not described adequately by the FWHM or FWTM. These can be caused by uniform bulk trapping [13], poor charge collection in partially dead windows on the detector [4], or nonideal responses in the amplifying chain.

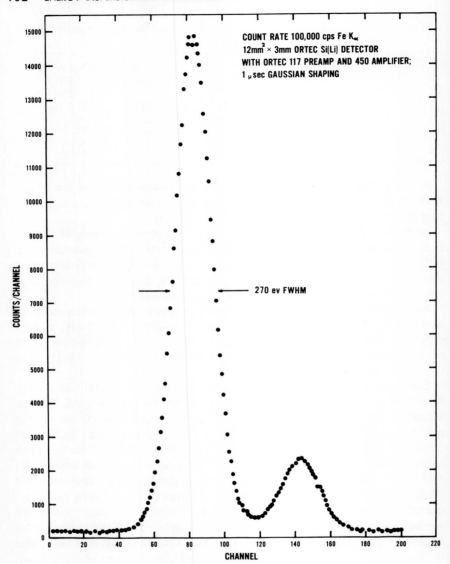

FIG. 8—*Good energy resolution at high count rate.*

Many of these second order spectral distortions show up very quickly in the peak-to-tail ratio or in the peak-to-valley ratio between the $K\alpha$ and $K\beta$. The effects of a partially dead window on a germanium detector are shown in Fig. 9, which compares two germanium detectors—one of which has a significant window problem (that is, an excessive low-energy tail). When comparing the tailing problem in different systems (as in Fig. 9) it is important that all spectra have an identical energy scale since the subjective

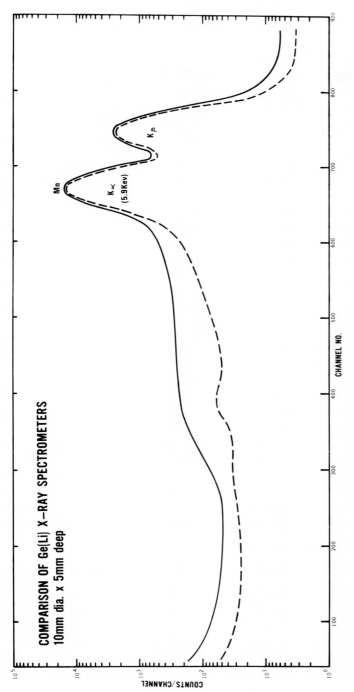

FIG. 9—*Low energy tail resulting from "partially dead" region at the entrance window on a Ge(Li) X-ray spectrometer.*

appearance of the tailing problem is very sensitive to the resolution width of the peaks.

The peak-to-tail ratio for a given peak (for example, Fe-55) is defined as the ratio of the intensity (counts/channel) at the center of that peak to the intensity (average counts/channel) at some specified lower energy. A convenient lower energy is that of the $K\alpha$-line of the next element down in the Periodic Table. Radioactive sources are the preferred X-ray source for measuring peak-to-tail ratios, since, if probably selected, they present the least problem in correcting for background. For example, if the X-rays are produced by electron excitation, the background intensity and shape will depend on specimen composition, exciting voltage, geometry, etc. Consequently, electron, X-ray, or charged particle excitation techniques should not be used for quantitative comparison of peak-to-tail performance unless all of the comparisons are done under identical conditions.

The peak-to-valley ratio (for example, for Fe-55) is defined as the ratio of the intensity (counts/channel) at the center of the manganese $K\alpha$-peak to the minimum intensity in the valley between the $K\alpha$ and $K\beta$-peaks.

Pileup Rejecion

Section CD of Fig. 4 can be moved to higher counting rates, and many of the undesirable features of region BC can be reduced (at a given *input* counting rate) by the use of pileup rejection circuits which prevent distorted (pulse-on-pulse or pulse-on-tail) events from being recorded in the MCA. Although this technique can be very useful in certain applications, its effectiveness can be easily overstated if one fails to recognize the large attendant loss in efficiency and the associated uncertainty in the counting losses (efficiency loss) as a function of count rate. Clearly, in most applications a 10^5 cps input rate which is recorded (stored in the MCA) at only 25 percent efficiency is less impressive than 5×10^4 cps input rate which is recorded at greater than 90 percent efficiency and the same energy resolution. Consequently, any performance results which are dependent on the use of pileup rejection should specify enough information about the data accumulation technique to allow one to estimate the rate of actual data storage as well as the resulting uncertainty in the real counting rate.

Radiation Detection Efficiency

The energy dependence of full energy peak efficiency of a semiconductor X-ray energy spectrometer is established by its geometrical size and shape, the presence of windows or dead layers on the detector and its container, charge trapping problems within the detector, the Z of the detector material, and (particularly in the case of germanium) the existence of escape phenomena near the absorption edge of germanium or silicon.

The electronic noise level in these systems has been reduced to the point where dead layer and window thickness problems are major contributors to the useful low-energy limit. Systems with one mil thick beryllium windows are now readily available, and even thinner windows are a distinct future possibility. However, there are good indications that dead layers on the detector itself are now sometimes the dominant problem. Consequently, the statement that a detector has a one mil beryllium window cannot be taken as a guarantee of good efficiency at low energies. In some applications, the useful low-energy limit will improve significantly when the manufacturers of X-ray equipment (electron probe microanalyzers, scanning electron microscopes, etc.) recognize that X-ray energy analysis is so essential that they must provide the access area, high quality clean vacuum, and safety interlocks which will eliminate the need for the isolating window. The remaining problem of background from low-energy electrons probably can be eliminated by magnetic and electrostatic deflection techniques which are used routinely in the nuclear physics field.

Typical graphs of efficiency versus energy are illustrated in Fig. 10. The low-energy rolloff is established by absorption in the window on the detector container (cryostat) and in any dead layer on the detector entrance surface. In addition, any partially dead layer near the entrance window may contribute to the low-energy rolloff in efficiency and may also result in low-energy background counts (that is, a low-energy tail on a monoenergetic X-ray peak). Even the metal electrode on the face of the detector can act as a partially dead layer since hot electrons in the metal can escape into the semiconductor.

Variations of full-energy peak efficiency as a function of energy also are caused by sharp changes in the absorption coefficient near the K_{abs} edge.

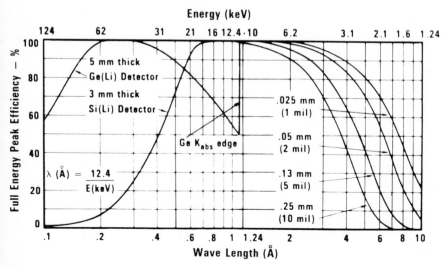

FIG. 10—*Full energy peak efficiency versus energy for several X-ray spectrometers.*

Since this sudden increase in cross section results in a high probability of interaction near the entrance window, dead or partially dead layers adjacent to the entrance window can cause serious drops in the full-energy peak efficiency at energies above and adjacent to the K_{abs} edge.

The high-energy rolloff in efficiency is established by the sensitive depth, sensitive area, and surface to volume ratio of the detector. The efficiency at high energies, as well as the spectral quality (low-energy tailing), also can be affected by the existence of a partially dead layer near the back contact to the sensitive region, poor charge collection throughout the sensitive region, or by isolated partially dead regions within the sensitive volume.

Although a number of methods for measuring the low-energy rolloff in efficiency have been used in the past [13], none of these seems to represent a realistic routine standard method which is practical for all users and manufacturers. One method for obtaining a relative measurement of the "window thickness" is to use a radioactive source with at least two lines whose relative intensities are well known. If one of these lines is located at an energy where the detection efficiency is near unity and the other at an energy which is well into the roll-off region, the relative observed intensity of these lines is an index of the window problem. Although the 14 and 6.4-keV lines from Co-57 have been used for this purpose, this is not a very satisfactory solution to the problem for either silicon or germanium. The energy of the 6.4 lines is much too high to be of interest for the very thin dead layers on state-of-the-art silicon detectors. In addition, the 14-keV line is in the region of the germanium response curve (above the K_{abs} edge) which can be significantly perturbed by the existence of germanium dead layers on the detectors. Since dead layer on the germanium window would attenuate both the 14 and the 6.4-keV lines, a measurement of their relative intensities is not a particularly sensitive indication of dead layer thickness. Electron beam excitation of fluorapatite ($Ca_{10}F_2[PO_4]_6$) to produce a set of X-ray lines (0.53-keV oxygen $K\alpha$, 0.69-keV fluorine $K\alpha$, 2.1-keV phosphorus $K\alpha$, 4.0-keV calcium $K\alpha$) of constant relative intensity also has been used as a sensitive relative measurement of the window thickness on thin window Si(Li) systems.[4,5] The practicality of using radioactive source excitation as a standard for window thickness measurements is currently under investigation. An ideal source for this application would have a reference line above 8 or 9 keV (above the energy where the low-energy rolloff on thick window systems might start) and below 11 keV (to avoid the germanium K_{abs} edge).

Pulse Pileup and System Paralysis

Peak width broadening by pulse-on-tail pileup is only one of the undesirable results which occur when the total pulse width is not very small

---

[4] Russ, J. C. personal communication.
[5] The Fluorapatite was obtained from Biodynamics Research Corp., Rockville, Md.

compared to the mean time interval between events. When pulse-on-pulse pileup in the shaping amplifier occurs, *two* pulses are moved from the energy location at which they belong to somewhere else in the energy spectrum (that is, into a false "sum peak" or into background). The resulting "system paralysis" is illustrated in Fig. 11 which shows how the output counting rate (in the manganese line) varies with the actual input rate of manganese X-rays [15]. Since the magnitude of the counting loss increases rapidly with increasing pulse-shape time constants, this illustrates another reason why using very long amplifier time constants to obtain better resolution (lower FET noise contribution) is detrimental to the count-rate capability of the system. Figure 11 also illustrates why it is important to use the actual output rate in the peak when quoting the count-rate performance characteristics (for example, resolution versus count rate) of a spectrometer.

Pulse Height Linearity

Electronic Linearity by the Pulser Method

An ideal pulse amplifier system produces an output pulse with an amplitude exactly proportional to that of the input pulse. Deviation from exact proportionality is described as nonlinearity. Integral nonlinearity L_I is defined as the maximum deviation from linearity expressed as a percentage of the manufacturer's specified maximum linear output. The measurement can be carried out with the pulser and multichannel analyzer hookup described previously for measurement of system noise. In order to include all of the amplifying chain in the measurement, the pulser must be coupled to the *input* of the preamplifier.

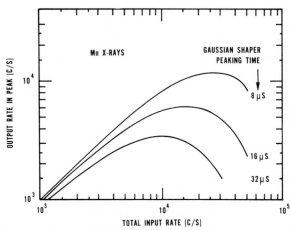

FIG. 11—*Plot of output counting rate in peak as a function of total input counting rate of the 5.9 keV X-rays for different Gaussian shaper peaking times. (Data courtesy of Landis et al, Ref 15).*

The measurement is carried out by setting the main amplifier gain to cover the energy region of interest and adjusting the pulse generator amplitude and the attenuator setting until the amplifier output equals the maximum linear output $V_{o(\max)}$, specified by the manufacturer. A pulser peak is accumulated in the MCA in channel $N_{(\max)}$ corresponding to amplifier output $V_{o(\max)}$. The pulser output amplitude V_P is reduced in increments (leaving all other settings fixed), and a pulser peak is accumulated in the MCA for each pulser setting.   The channel number N corresponding to the maximum of the mono-energetic line is plotted as a function of V_P. For a given choice of amplifier settings and after correction for any nonlinearities in the pulser and MCA, the integral nonlinearity L_I is given by

$$L_I = \frac{100 \, |\Delta N|_{(\max)}}{N_{(\max)}} \text{, (in percent)}. \dots \dots \dots \dots \dots (9)$$

where ΔN is the difference between the observed N (for a given V_P) and the N that would have been predicted from a straight line joining $N_{(\max)}$ and the origin, and where $(\Delta N)_{(\max)}$ is the largest value of ΔN anywhere between $N_{(\max)}$ and the origin. When characterized by a single unqualified number the L_I shall not exceed that number at any amplifier settings.

It should be noted that the linearity of both the amplifier and the MCA can be sensitive to the type of pulse shaping and the pulse shaping time constants. Consequently, the same pulse shape should be used for testing both the linearity of the pulser-MCA combination and the X-ray energy spectrometer system.

X-ray Response Linearity

Nonlinearities in the response of the detector to X-rays can produce a nonlinear system response. Previous extensive experience with semiconductor detectors used for nuclear physics research indicates that this probably will not be a significant problem at high X-ray energies. However, preliminary evidence [4,13,14] suggests that the detector response may be nonlinear in the low-energy region where partially dead layers at the surface (entrance window) of the detector and electron escape problems play a dominant role. Charge loss from these mechanisms can not only broaden the resolution as mentioned previously, but also can reduce the most probable value of the pulse height. In addition, there is no firm evidence that ϵ is independent of energy at these low energies. Good linearity data on a significant sampling of detectors are needed badly to clarify this issue.

X-ray response linearity measurements are made in the same way as the electronic linearity test described previously, except that the incremental series of pulser settings is replaced with an incremental series of characteristic X-ray lines spanning the energy region of interest. Since charge trapping problems are involved, it should be anticipated that this measurement also may be sensitive to the amplifier passband time constants.

Detector Material Selection: Germanium Versus Silicon

The two semiconductor detector materials which are presently available for high resolution X-ray energy spectrometers are germanium and silicon. The relative advantages and disadvantages of each are tabulated below:

Germanium

Advantages—

1. Higher Z which results in better efficiency at high energies (germanium is reasonably efficient up to 100 keV and usable at energies > 1 MeV).
2. The smaller value of ϵ reduces the relative importance of preamplifier noise.
3. The smaller experimentally verified upper limit on the ϵF product suggests a possible inherent resolution advantage.

Disadvantages—

1. The lithium compensation of germanium devices is unstable at room temperature. (Recent improvements in ultrahigh purity germanium technology could eliminate this problem).
2. Higher efficiency at high energy produces more background problems from high energy sources.
3. Higher efficiency at high energy increases the probability of amplifier overload.
4. An equivalent linear thickness for the dead layer (entrance window) corresponds to a much larger efficiency loss at low energy.
5. Higher Z results in increased spectral complication by escape peak problems and a more complex relation between efficiency and energy.
6. Smaller band-gap increases the potential detector noise problems and decreases the maximum detector operating temperature.

Silicon

Advantages—

1. Better lithium stability and lower detector noise at higher temperatures.
2. Relative freedom from escape peak complications in complex spectra except at low energies.
3. Thin dead layers.
4. Less sensitive to high-energy background problems.

Disadvantages—

1. Low efficiency at high energy (> 30 keV).
2. A larger experimentally verified upper limit on the ϵF product.

The above tabulation suggests that germanium (because of its smaller ϵF product) should have better energy resolution capabilities than silicon. However, to date germanium detectors have failed to equal the resolution capabilities of silicon devices. Part of this discrepancy can be identified with practical problems (associated with the smaller band-gap in germanium) which probably will be solved eventually. However, in the meantime, germanium detectors are the appropriate choice only for high energies where the efficiency of silicon devices is unacceptable. Even if germanium eventually should provide better resolution, its relative position may still be handicapped by the fact that elimination of dead layers on germanium is a more difficult problem (because of its higher Z). Spectra taken with germanium detectors also will bear the handicap of complicating escape peaks in the X-ray energy region of typical interest and a complicated efficiency versus energy curve caused by the germanium absorption edge which is located at approximately 11 keV.

Other Design Considerations and Compatibility with Associated Equipment

Microphonics, Secondary Fluorescence, LN$_2$ Loss Rate

Microphonic noise generation may be a serious source of resolution broadening. In cryogenic systems, mechanical vibration of the input circuit components with respect to surfaces at different potentials are induced by bubbling of liquid nitrogen (LN), turbulence from automatic nitrogen transfer systems, vibration of other equipment in contact with the detector cryostat, and environmental noise. To generate a microphonic signal with the amplitude of 10 keV (silicon), a change of capacitance between the FET gate and the high voltage (detector bias 1000 V) of only 5×10^{-7} pF is required.[3] Since the typical stray capacitances are of the order of 1 pF, relative mechanical motions (within the frequency pass-band of the shaping amplifier) of approximately 10^{-7} are significant. Since most of the vibration problems are at relatively low frequency, much of the microphonic noise is filtered out by the limited amplifier pass-band. Consequently, microphonics problems are usually more serious at long shaping time constants. Since this problem usually involves the excitation of internal mechanical resonances by internal or external environmental mechanical vibrations, there is no straightforward way to establish a test standard for microphonics. However, the problem is very sensitive to the internal design of the spectrometer, and therefore the user should be aware of the problem.

Secondary fluorescence of metal parts of the cryostat, detector and preamp mount, and the metal electrodes on the detector can cause spectral distortion in the form of artifact peaks in the spectrum.

Most semiconductor X-ray energy analyzers use liquid nitrogen as a source of cooling for the detector and preamplifier. The efficiency of the cryostat design is directly reflected in the LN consumption rate. Excessive LN consumption or frequent LN filling aggravates the microphonics problem, increases the operation cost and inconvenience, and increases the chance of accidental warm-up. This performance parameter usually is covered by specifying the capacity of the LN reservoir and the minimum guaranteed holding time.

Solid Angle and Takeoff Angle

The solid angle between the source of fluorescent X-rays to be analyzed and the detector establishes the geometrical counting efficiency of an energy spectrometer. Thus, both detector area and the source to detector distance are important for establishing what the counting rate will be.

Takeoff angle (angle between the surface of specimen which is emitting the X-rays and a line from the source of X-rays to the center of the detector) is also an important variable since different takeoff angles require different quantitative analysis corrections for selfabsorption and other matrix effects.

Therefore, when an X-ray energy analyzer is intended for use with a specific X-ray machine (probe, scanning electron microscope, fluorescence analyzer, etc.), the takeoff angle and source to detector distance are important specifications.

Multichannel Analyzers and Comptuer Interfaces

The characteristics of the multichannel analyzer (MCA) (or the analog-to-digital converter (ADC) and computer system) can directly influence the performance of the energy analysis system. For example, when a semiconductor energy analyzer is optimized for high count rate, its count-rate capability can exceed the count rate capability of presently available multichannel analyzers. In addition, in order to maintain good energy resolution as count rate is increased, it is necessary to either d-c couple or provide good baseline restoration up to the ADC of the MCA. Therefore, if the ADC does not provide for true d-c coupling, the overall count-rate performance of the system will suffer.

To optimize the spectrometer for widely different count-rate situations, it will be necessary to operate at different amplifier time constants which may vary by as much as several orders of magnitude. Consequently, analyzers which will work well over only a narrow range of input pulse shapes will handicap the operator's ability to optimize the system performance unless an appropriate pulse shape interfacing circuit (for example, a gated stretcher) is provided.

The linear gates on some ADC's are paralyzable, a factor which may considerably complicate quantitative analysis at high count rates.

Acknowledgments

The author is indebted to Emanuel Elad, Rex Trammell, and other members of the ORTEC staff for critical review of the manuscript. The data used in some of the illustrations were obtained by Emanuel Elad and Richard Dyer. Dale Gedcke calculated the results shown in Table 2.

References

[1] Copies of IEEE No. 300 and IEEE No. 301 may be obtained from: Standards Department, The Institute of Electrical and Electronic Engineers, 345 East 47 Street, New York, N. Y., 10017.

[2] Elad, Emanuel, *Nuclear Instruments and Methods*, Vol. 37, 1965, p. 327.

[3] Heath, R. L., *Semiconductor Nuclear-Particle Detectors and Circuits*, Brown, L. et al, eds., NAS-NRC 1593, 1969, p. 247.

[4] Walter, F. J., to be published *Transaction on Nuclear Science*, Institute of Electrical and Electronic Engineers, Proceedings of the 12th Scintillation and Semiconductor Counter Symposium, March 1970.

[5] Pehl, R. H. et al, *Semiconductor Nuclear-Particle Detectors and Circuits*, Brown, L. et al, eds., NAS-NRC 1593, 1969.

[6] Walter, F. J. et al, *Semiconductor Nuclear-Particle Detectors and Circuits*, Brown, W. L. et al, eds., NAS-NRC 1593, 1969.

[7] Zulliger, H. R. and Aitken, D. W., to be published in *Transaction Nuclear Science*, Institute of Electrical and Electronic Engineers, Proceedings of the 12th Scintillation and Semiconductor Counter Symposium, March 1970.

[8] Goulding, F. S. et al, UCRL-19377 preprint, to be published in *Transactions on Nuclear Science*, Institute of Electrical and Electronic Engineers.

[9] Radeka, Veljko, to be published in *Transactions on Nuclear Science*, Institute of Electrical and Electronic Engineers, Proceedings of the 12th Scintillation and Semiconductor Counter Symposium, March 1970.

[10] Nowlin, C. H. and Blankenship, J. L., *Review of Scientific Instruments*, Vol. 36, 1965, p. 1830.

[11] Chase, R. L. and Poulo, L. R., *Transactions on Nuclear Science*, Institute of Electrical and Electronic Engineers, NS-14, No. 1, 1967, p. 83.

[12] Trammell, Rex and Walter, F. J., "The Effects of Carrier Trapping in Semiconductor Gamma-Ray Spectrometers," *Nuclear Instruments and Methods*, Vol. 76, 1969, p. 317.

[13] Palms, J. E. et al, *Transactions on Nuclear Science*, Institute of Electrical and Electronic Engineers, NS-16, No. 1, 1969, p. 36.

[14] Zulliger, H. R. et al, *Transactions on Nuclear Science*, Institute of Electrical and Electronic Engineers, NS-16, No. 1, 1969, p. 47.

[15] Landis, D. A. et al, UCRL report UCRL-19796, June 1970, to be published in *Nuclear Instruments and Methods*.

A. O. Sandborg[1]

Energy Dispersion X-ray Analysis with Electron and Isotope Excitation

REFERENCE: Sandborg, A. O., **"Energy Dispersion X-ray Analysis with Electron and Isotope Excitation,"** *Energy Dispersion X-ray Analysis: X-ray and Electron Probe Analysis, ASTM STP 485,* American Society for Testing and Materials, 1971, pp. 113–124.

ABSTRACT: The components of the energy dispersion X-ray analysis system are discussed, and the electronic noise of peak broadening due to each is identified. Data on count rates and sensitivity are listed for various elements to enable the prospective user to determine whether the technique is applicable to his specimens. Specific comparison is made between the use of electron beams, radioactive isotopes, and X-ray tubes to excite the characteristic X-ray lines of interest. The ability to distinguish small peaks when Bremsstrahlung (white radiation) is present due to the use of direct electron excitation, or when scattering of incident X-rays or gamma rays occurs is discussed.

KEY WORDS: dispersing, X-ray analysis, X-ray fluorescence, electron probes, scanning electron microscopy, transmission electron microscopy, electron beams, radioactive isotopes, X-ray tubes, X-ray spectra, electronic noise, semiconductor devices, silicon, field effect transistors

Energy dispersion X-ray analysis has become a useful technique in the fields of X-ray fluorescence, electron microprobe analysis, scanning electron microscopy (SEM), and transmission electron microscopy (TEM). It always has had the advantage of speed of analysis because of its ability to analyze all elements at once and its high detection efficiency. In recent years, the combination of silicon radiation detectors with field-effect transistor (FET) preamplifiers and the optimization of that combination has yielded the resolution necessary to allow separation of $K\alpha$-lines for all elements above atomic number 12.

A prospective user of this technique is often quite unfamiliar with either its total capabilities or, in some cases, with X-ray analysis itself. Manufacturers specify system resolution and detector area, but, in general, little data are given about actual performance. Thus, it is difficult to estimate the sensitivity of the technique for the elements of interest and the time

---

[1] Vice president of research, Nuclear Diodes, Inc., Prairie View, Ill. 60069.

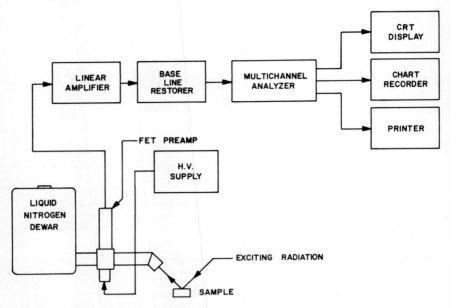

FIG. 1—*Block diagram of energy dispersive X-ray analysis system.*

necessary to obtain an analysis. The purpose of this paper is to present data that will allow the user to relate instrument specifications to expected results, and to permit comparison of analytical results obtainable using electron and isotope excitation.

Description of System

Figure 1 illustrates the components of an energy dispersion X-ray analysis system, as described in greater detail in the paper by Fitzgerald and Gantzell.[2] The first three segments of the system are each important, and even more important is their integration. These are the detector, the field-effect transistor (FET) preamplifier, and their cryogenic enclosure. The enclosure must be designed to cool the detector and the FET to their indivdual optimum temperatures, to damp any significant microphonic noise, and to present the detector to the X-rays to be detected. The detector is a planar, lithium drifted silicon diode generally 3 mm in thickness and of a diameter from 3.5 to 12 mm. It must have very low leakage current ($> 10^{-12}$), low capacitance, and good charge collection when cooled to cryogenic temperatures. The X-rays enter the detector through a thin (100 to 200 Å) gold layer into the intrinsic silicon. This gold layer absorbs some X-rays, of course, and there apparently is an additional absorption in the silicon caused by poor charge collection immediately under the gold.

[2] See p. 3.

The FET is a silicon junction FET of a type chosen to have high G_M and low capacitance, but also selected to have low noise when cooled to 120 to 150 K. The preamplifier is usually of a charge sensitive design with some feedback to stabilize the operating point of the FET. The various methods of providing this feedback are discussed below.

The intermediate electronics consist of a high voltage supply that biases the detector to collect the charge created by the X-rays, an amplifier which amplifies the pulses from the preamp and shapes them to give optimum signal-to-noise ratio before pulse height analysis, and a baseline restorer, which prevents d-c level baseline shifts at varying count rates before the signals enter the analyzer.

Lastly, we have a multichannel analyzer. Little space will be devoted here to its requirements; however, it should be d-c coupled into its analog-to-digital converter, accept a wide range of pulse shapes including the wide pulses which give optimum resolution, and finally be organized to allow its user to quickly identify the elements whose peaks appear in the spectrum.

Discussion of System Parameters Which Affect Resolution

We consider now those factors which affect the resolution of an energy dispersive system.

The great improvements in resolution in these systems primarily have been made by reducing the electronic noise of the system, which originates chiefly in the input stage of the preamplifier. This is the input FET, and the noise is due to the FET itself and the components connected to its gate. The noise can be described as the sum of two parts, one constant and one proportional to the capacitance added by the other components. The value for the FET alone is difficult to measure directly since one cannot successfully operate a preamplifier without any components connected to the gate of the FET. From work recently done [1],[3] however, the noise intercept at zero added capacitance can be less than 80 eV. The proportionality constant for the FET's normally used is of the order of 30 eV per pF of added capacitance. The components attached to the gate of the FET are the detector and the feedback components of the preamp. The detector adds a noise contribution due to its leakage current and its capacitance. A leakage current of about 10^{-13} A adds a noise of less than 10 eV, while the capacitance of from 1 to 3 pF adds 30 to 90 eV. In the feedback loop, a capacitor almost always is added; its 0.2 pF capacitance adds 6 eV. In a resistive feedback system, the resistor contributes, at best, about 60 eV [2], and in fact must be carefully selected to approach that value. In an optical feedback system [3,4], light is used to create a current in the gate-to-drain junction of the FET equivalent to the detector current, so that the noise due to the detector-FET light current combination is equal to 1.414

[3] The italic numbers in brackets refer to the list of references appended to this paper.

times the noise of the detector current alone. Some feedback method is required at all but the lowest count rates to maintain the proper operating voltages on the FET. To eliminate the noise due to feedback, a system has been developed by Goulding [1] which allows the system to operate without feedback until the FET has almost turned off. This fact is sensed by a discriminator, and optical feedback is applied to the FET to restore it to its original operating point. During the time of feedback application, about 10 μs, the system is turned off. As shown in Fig. 2, this technique gives excellent count-rate performance. The upper limits in count-rate performance are due to baseline shifts which the baseline restorer can no longer restore.

When an X-ray enters the detector and interacts, each hole-electron pair produced takes 3.8 eV. Resolution is affected by the statistical nature of the production of these ionizations. If the process obeyed Poisson statistics, the resolution would be proportional to $\sqrt{N}$, the number of pairs. In practice, however, the observed resolution is much better than this, equaling $\sqrt{FN}$, where F, the Fano factor, is less than one. There has been a great deal of debate about the Fano factor [5], since as detectors have improved, measured values have been decreasing, and the real value of it is still not known. Theoretically, for silicon, it can be as low as 0.05. Even though the true value is not known, an effective value can be measured by the relation:

$$F_{EFF} = \frac{R_I{}^2 - R_E{}^2}{(2.355)^2 \epsilon E_x}$$

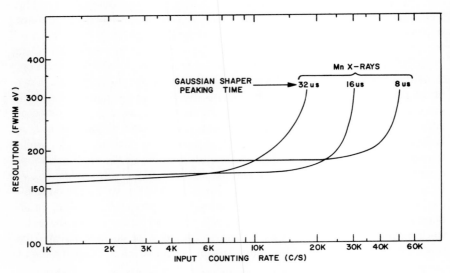

FIG. 2—*Count-rate performance of pulsed d-c feedback system.*

where:

R_T = system resolution, FWHM, eV;
R_E = electronic resolution FWHM, eV;
ϵ = 3.8 eV;
E_x = energy of X-ray, eV.

This effective value, of course, lumps together all factors which affect detector resolution (such as poor charge collection) in addition to real statistical effects. Most detectors of reasonable quality exhibit an effective Fano factor of 0.13 or less.

Discussion of System Parameters Which Affect Efficiency

The sensitivity and hence the usefulness of the energy spectrometer depend strongly on its efficiency. The detector is a right circular cylinder whose thickness (typically 3 mm) determines its ability to stop high-energy X-rays, while absorption in cryostat windows and dead layers on the detector determine the response at low energies. Figure 3 shows the spectral response of a 3-mm-thick device with various window thicknesses of beryllium.

Figure 3 does not include any effects of absorption of X-rays in the surface layers of the detector. These have been discussed by Aitken and Woo.[4] Although they suggest that absorption in this thin layer of material is not serious, the fact remains that the charge collection near the surface of a detector can be less than ideal. Gross trapping effects rarely are seen with good quality silicon. With poor silicon, however, tailing on the low-energy side of a peak is observed. Generally a peak shape that is near Gaussian down to full width at one-tenth maximum (FWTM) is evidence of good charge collection.

Parameters Which Affect Peak-to-Background and Sensitivity

In selecting a detector system, the primary specification is commonly resolution. Although this is a very important factor, others also must be considered. The first important result we are trying to achieve is element separability; that is, the ability to resolve two or more X-ray lines from different elements which lie close in energy to one another. This ability can be inferred from the system resolution. If the full width at half maximum (FWHM) is about two thirds the energy spacing between two roughly equal peaks, one will be able to observe a valley between them; almost complete separation is achieved when the FWTM is about two thirds the spacing between the two peaks.

Secondly, we are interested in the minimum amount that can be detected of a given element. As has been pointed out [6,7],[5] the minimum

---

[4] See p. 36.
[5] See p. 154.

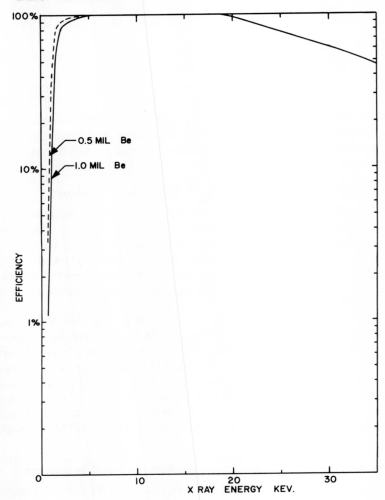

FIG. 3—*Spectral response of Si(Li) detector.*

detectable limit is proportional to $1/\sqrt{1_o^2/1_B}$. Maximizing the product of $1_o/1_B \times 1_o$ will give the best sensitivity. $1_o/1_B$ is the integrated peak to background ratio; its value depends upon the method of excitation used. 1_o is the integrated count rate, which can be increased by moving the detector closer to the source of X-rays, to collect a larger solid angle, or by increasing the size of the detector. The former is preferable since it has no effect on resolution, but the latter generally broadens the resolution due to the added capacitance of the larger area detector. The resolution increase in going from 10 to 100 mm² is less than a factor of two, indicating that the largest detector would have a much better sensitivity. However, that resolution increase may often prevent separation of adjacent peaks from light elements.

Mention of the sources of background is appropriate at this point. With electron excitation there is a considerable background because of continuous X-rays (Bremsstrahlung) produced in the specimen along with the characteristic X-rays. This white radiation has a maximum energy equal to that of the electron beam voltage, and a maximum intensity at an energy about two thirds of the electron beam voltage. The intensity varies directly with the atomic number of the specimen and as the square of the electron beam voltage. In general, peak to background increases as one increases beam voltage due to more efficient excitation of the characteristic radiation. Eventually, peak to background may decrease[5] if characteristic excitation efficiency drops due to internal absorption; in practice, this generally is observed only for the lightest elements.

With X-ray excitation, the background comes from Compton and coherent scattering of the primary X-rays. In the case of an X-ray tube source, this will be scattering of the continuum from the tube and will appear as a continuum. However, the magnitude will be for less than that from direct electron excitation. In the case of radioisotopic excitation where we have essentially monoenergetic radiation, the scattered radiation will be limited to a given energy range, and, by proper selection of the primary energy, one can obtain very little background for some elements.

Finally, the detector itself contributes background. Each peak that is produced by the detector has a low-energy background associated with it due to incomplete charge collection, which then becomes background for a lower energy X-ray. (This is demonstrated in the section on radioisotopic excitation.)

Relation of System Parameters to Performance with Electron Excitation

Elemental standards were excited with an electron beam in a Jeolco Model JSMU-3 scanning electron microscope, and the X-rays detected with a Nuclear Diodes Model XS-12-200-JSMU-3 EDAX system. Figure 4 illustrates the apparatus. The area of the detector is 12.5 mm^2, and its thickness was 3 mm; a 1 mil beryllium window was used on the cryostat; and the detector was located 2.2 in. from the specimen. The data in Table 1 show peak-to-background ratios that are not as high as those obtained with X-ray excitation for the reasons described above.

The count rate is high for light elements, decreasing for the lightest elements due to the window absorption and decreasing again for heavy elements due to inefficient excitation. Count rates and peak-to-background ratios vary with beam voltage as shown by Russ.[5] In general, the highest accelerating voltages yield the highest sensitivity.

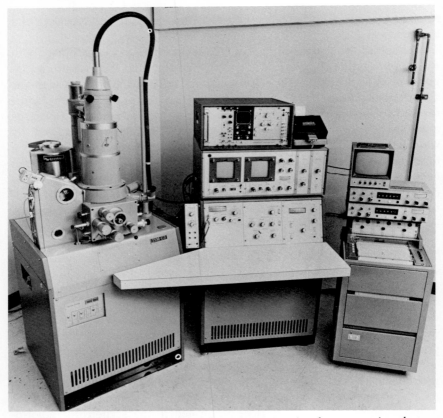

FIG. 4—*Energy dispersion X-ray analysis system interfaced to a scanning electron microscope.*

TABLE 1—*Elements excited with electron beam*[a]

Atomic Number	Symbol	Count Rate, cps/pA	P/B	I_o/I_b	I_o^2/I_b, per pA	FWHM, eV
13..........	Al	4.27	81	34.7	148	150
22..........	Ti	12	77	31.4	377	181
26..........	Fe	10.4	77	33	343	191
42..........	Mo	2.6	21	10.4	27	308

[a] Beam voltage, 40 kV for Al, 50 kV for Ti, Fe, and Mo. Jeolco SEM, Model JSMU-3.

Relation of System Parameters to Performance for Radioisotope Excitation

To illustrate some of the capabilities of EDAX when used with radio-isotopic excitation, a 30-mm² area 3-mm-thick silicon detector was used to detect the X-rays from elemental standards excited with radioactive sources. The specimen were all 1 in. in diameter, "infinite thickness," and were located 2 in. from the detector, which had a 1/2 mil beryllium window. Figure 5 illustrates the apparatus, and Fig. 6 shows a schematic of its operation. The sources were located 1.5 in. from the specimen, and the angle between the detector to specimen and specimen to source was 45 deg. Elements up to atomic number 24 were excited with a 10-mCi Fe⁵⁵ source, which emits manganese X-rays; for these specimens a vacuum was used to eliminate absorption of the X-rays in air. The data in Table 2 show that count rate increases rapidly with atomic number due to the increasingly more efficient excitation as the absorption edge of the excited element approaches the energy of the source emitted X-rays. (Chromium is excited by only the $K\beta$'s of manganese which accounts for its lower count rate.) Peak-to-background also increases rapidly with atomic number. Poor sensitivity will be obtained on the lightest elements under these excitation conditions due to the decrease in fluorescent yield and absorption in the window of the detection system.

TABLE 2—*Excitation with Fe⁵⁵ 10 mCi.*

Atomic Number	Symbol	Count Rate, cps	P/B	I_o/I_b	$I_o{}^2/I_b$	FWHM, eV
12	Mg	8.3	36	18.5	153	190
13	Al	17.3	63	33	572	200
14	Si	40	200	100	4 000	205
22	Ti	450	410	175	79 000	222
24	Cr	88	76	34	2 980	232

Table 3 shows data taken with a Cd¹⁰⁹ source whose primary radiation consists of silver X-rays. The source used was a 0.8 mCi; however, 10-mCi Cd¹⁰⁹ sources are available and would give proportionately higher count rates. Peak-to-background ratios are high, so that with a stronger source excellent sensitivities are possible.

TABLE 3—*Excitation with Cd¹⁰⁹ 0.8 mCi.*

Atomic Number	Symbol	Count Rate, cps	P/B	I_o/I_b	$I_o{}^2/I_b$	FWHM, eV
26	Fe	11.2	182	82	920	231
28	Ni	19	333	141	2 680	235
29	Cu	25.4	562	223	5 700	246
32	Ge	37.0	311	132	4 880	259
42	Mo	81.6	351	157	12 800	320

For elements above atomic number 24, an Am²⁴¹ source was used, whose primary energy is a 59.6-keV gamma ray. The data in Table 3 show low counting rates and poor peak-to-background ratios for iron, nitrogen, and

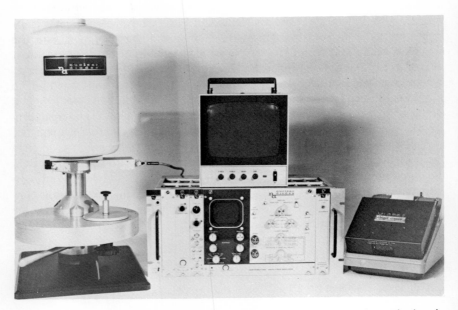

FIG. 5—*Energy dispersion X-ray analysis system for radioisotopic excitation in vacuo.*

copper. The low counting rates are due to poor excitation because of the large difference in energy between the absorption edge of the specimen, and the exciting energy. Background is high because of a large amount of Compton scattering in the specimen which yields a large backscatter peak. This peak has an energy of about 50 keV; but a sizeable fraction of these backscattered gammas yield Compton events in the silicon detector itself, which fall in the range of these three elements. Germanium falls above this range and yields a much higher peak-to-background ratio even though the count rate has only increased by a factor of two over copper. At high atomic numbers, peak to background decreases somewhat due to broadening in resolution from statistical processes in the detector and the splitting of the $K\alpha_1$ and $K\alpha_2$ peaks.

With the exception of those elements excited by Am^{241} where Compton scattering contributes to background, one of the major sources of background is the tail on the low-energy side of each peak. The relative amount of this tail, due as mentioned to incomplete charge collection in the detector, can vary from detector to detector. If the tail is large, the sensitivity will be decreased for the lower in energy of two or more elements being simultaneously analyzed. This points up the need for considering the magnitude of the tail or peak-to-tail ratio, and also the need for better understanding of the reasons for its existence.

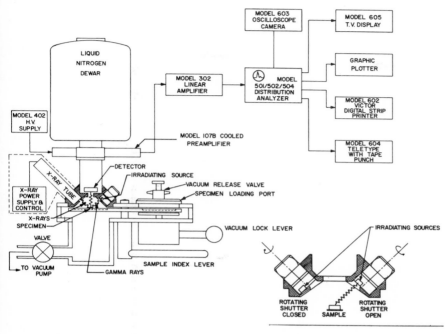

FIG. 6—*Block diagram of EDAX system for radioisotopic excitation.*

Discussion

The data in Tables 1, 2, 3, and 4 allow comparison of the count rates and sensitivities obtainable with electron and radioisotopic excitation. Since the beam currents normally used in an electron microscope or microprobe can range from 10 picoamperes (pA) to 100 nanoamperes (nA), it is evident that the count rates obtainable in the electron excited case are much higher than practical with radioisotopic excitation. The count rates with isotopic excitation can be increased by using a more sophisticated geomtery.

TABLE 4—*Excitation with Am^{241} 10 mCi.*

Atomic Number	Symbol	Count Rate, cps	P/B	I_o/I_b	I_o^2/I_b	FWHM, eV
26	Fe	22	25	12	264	231
28	Ni	37	52	26	960	235
29	Cu	44.5	74	26	1 160	246
32	Ge	92	332	101	9 300	272
42	Mo	372	335	161	60 000	337
49	In	601	229	97	58 100	444
50	Sn	670	213	94	63 000	465
51	Sb	698	272	113	78 800	503

However, for the very lightest elements (silicon and below), electron excitation remains superior to excitation with Fe^{55}. It also appears that with a few exceptions, electron excitation is generally more sensitive to elements below atomic number 30. Isotopic excitation, on the other hand, is much more sensitive to the elements higher than atomic number 30. These data help to justify the trend to smaller, higher resolution detectors for electron microscope and microprobe applications, where count rates are adequate and light elements are excited easily. Since sensitivity is high for the light elements, it is important that the resolution be good enough to separate adjacent light elements. Even though a larger, poorer resolution detector would appear to give higher sensitivities, the loss of light element separability would generally rule against the larger detector.

In contrast, radioisotopic excitation gives low count rates for light elements and higher count rates for heavier elements. A larger area detector is of advantage since it yields higher sensitivity. Furthermore, the resolution at the higher energies of interest is beginning to be controlled by detector effects and the increase in separation between the $K\alpha_1$ and $K\alpha_2$-lines rather than by the electronic contribution to resolution from the added capacitance of a larger detector.

Data on excitation from an X-ray tube were not available at the time of this writing; however, it would appear that this form of excitation would have the advantages of both electron and radioisotopic excitation. Count rate should be high, background low, and a wide range of elements should be excited. Small, high resolution detectors would again be preferable.

References

[1] F. S. Goulding, to be published.
[2] McKensie, Kern, "Noise Studies of Ceramic Encapsulated Junction Field Effect Transistors," *Transactions on Nuclear Science,* Institute of Electrical and Electronic Engineers. Vol. NS-17, No. 3, June 1970.
[3] Goulding, F. S., Walton, J., and Malone, D. F., "An Optoelectronic Feedback Preamplifier for High Resolution Spectroscopy," Lawrence Radiation Laboratory, UCRL-18698.
[4] Goulding, F. S., Walton, J., and Pehl, R. H., "Recent Results on the Optoelectronic Feedback Preamplifier," *Transactions on Nuclear Science,* Institute of Electrical and Electronic Engineers, Vol. NS-17, No. 3, June 1970.
[5] Aitken and Zulliger, "Fano Factor Fact and Fallacy," *Transactions on Nuclear Science,* Institute of Electrical and Electronic Engineers, Vol. NS-17, No. 3, June 1970.
[6] E. Lifshin, Technical Information Series Report 69-C-346, General Electric Co., Oct. 1969.
[7] Sutfin, L. V. and Ogilvie, R. E., "A Comparison of X-ray Analysis Techniques Available for Scanning Electron Microscopes," *Proceedings of the Third SEM Symposium,* Chicago, April 1970.

Garry Williams[1]

Role of Multichannel Analyzer in Data Handling

REFERENCE: Williams, Garry, **"Role of Multichannel Analyzer in Data Handling,"** *Energy Dispersion X-ray Analysis: X-ray and Electron Probe Analysis, ASTM STP 485,* American Society for Testing and Materials, 1971, pp. 125–139.

ABSTRACT: The multichannel analyzer role in X-ray spectrometer analysis has been taking on an ever increasing importance. Complete energy spectra observations are now possible from a single scan of a specimen with electron microscope.

The scope of this paper will be to interpret these spectral data in terms of accuracy, parameters, and usefulness to the X-ray spectrometer user. A set of general specifications will be outlined for the multichannel analyzer to cope with the ever changing requirements of the user and developments of the manufacturer.

With this aid it is the hope that the users of the multichannel analyzers will be able to incorporate units in their system that will perform the present and future functions of their applications.

KEY WORDS: X-ray spectrometers, analog computers, digital computers, electron microscopy, X-ray spectra, X-ray analysis, address scaler, analog to digital converter, analysis mode, base line restorer, binary coded decimal, coincidence input, conversion gain, differential linearity, digital overflow, digital zero offset, display mode, input signal, integral linearity, linear gate, lower level discriminator, magnetic core memory, memory, memory cycle, memory data scaler, multichannel analyzer, multichannel scaling, parallel entry memory, pulse amplitude to time converter, pulse height analyzer, readout mode, serial entry memory, single channel analyzer, stretcher capacitor, upper level discriminator, evaluation, tests

The multichannel analyzer (MCA) is a special purpose computer programmed for the function of analyzing random events. Therefore, it has become an important part of most nuclear, medical, and chemical experiments involved in analyzing particle energy from alpha, beta, gamma, and X-rays. The collected energy information in the analyzer are called spectra and appear as shown in Fig. 1. To further analyze the spectra, digital and analog functions are wired into the analyzer to perform arithmetic, integration, normalizing, and calibration operations. These operations have become routine qualitative analyzing features of the MCA and will be ex-

[1] Vice president, Northern Scientific, Inc., Middleton, Wis. 53562.

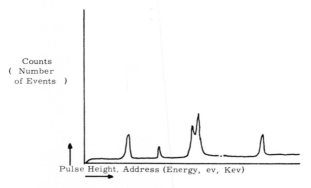

FIG. 1—*MCA display of X-ray energy spectra.*

amined during the discussion of the system (Fig. 2). The accuracy and flexibility of interpreting spectral data depend heavily on initial performance specifications set for the MCA. Therefore, the main function of this paper will be to outline standards for selecting an MCA in its role of data handling in a typical X-ray energy dispersive system.

The MCA system has logical block functions that will be used for discussion. These blocks consist of the analog-to-digital converter (ADC), memory unit, and associated peripherals (Fig. 2).

Analog-to-Digital Converter (ADC)

Principle of Operation

The ADC is the heart of the system. Through the ADC, X-ray detector output pulses are converted to a digital form for acceptance by the memory system, thus establishing the accuracy of the information to be stored in the analyzer's memory.

In the design growth of ADC many converting methods were tried for improving the converting time, linearity, and resolution. These methods varied from successive approximation to optical decoders.[2] From this search evolved two types of units; the industrial and research ADC. The industrial ADC's consist of digital voltmeters, temperature and pressure converters, and assembly line monitors, all having converting speed and resolution but lacking in linearity which is needed most in research systems. Therefore, with the research ADC evolved a unit that gave linearity, resolution, and moderate conversion speed.

The most popular converting method today for the research ADC is the pulse amplitude to time converter. The operating principle of the pulse amplitude to time converter is to charge a capacitor to a voltage (V_c) proportional to the input signal voltage (Fig. 3). With the aid of the diode or

[2] Chase, R., *Nuclear Pulse Spectrometry*, McGraw-Hill, New York, 1960, p. 81.

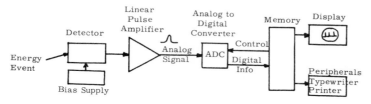

FIG. 2—*X-ray energy spectrometer components diagram.*

similar electronic component, the capacitor is charged on the positive portion of the input signal, and when the input signal peak is detected by the "comparator," the constant current source is enabled, linearly discharging the stretcher capacitor to zero. During the discharge time the crystal clock is enabled, and the output pulses are counted into a digital scaler. The final number of pulses in the scaler is the digital representation of the input analog pulse amplitude.

The detailed operation of the ADC is explained best by following an input signal through the logic block in Fig. 4. Before any input signal is accepted certain input requirements must be met to guarantee a high quality conversion. These requirements are:

1. The input signal amplitude must be larger than the lower level discriminator (LLD) which is set to be above the input noise or level of unwanted signals.

2. The input signal amplitude must be less than the upper level discriminator (ULD) setting to guarantee that the input pulse is in range of the ADC conversion capabilities and not an "overloaded" nonlinear amplifier pulse.

3. Internal logic from the ADC must be satisfied indicating that no internal events are being processed.

Along with these initial logic requirements, the input signal is "conditioned" for entry into the ADC. Therefore, there are internal threshold controls for rejecting input noise (besides LLD) and a zero level control for establishing the zero channel energy intercept.

A typical application of the zero level control is shown in Fig. 5. The top spectrum is an energy spectrum of 0 to 10 keV with channel zero being equal to 0 keV and the Nth channel equal to 10 keV. By adding a voltage bias or "offset" to the input signal (increasing the zero level control) the 10 keV point is moved to channel zero and the Nth channel becomes 20 keV. With some ADC's the zero level control can double the effective range of the system. For example, an ADC of 1024 channels with a 100 percent "zero offset," becomes a 2048 channel unit by doubling the input signal amplitude. *Note*—For final calibration of an ADC the zero control is very necessary for establishing the channel zero energy intercept.

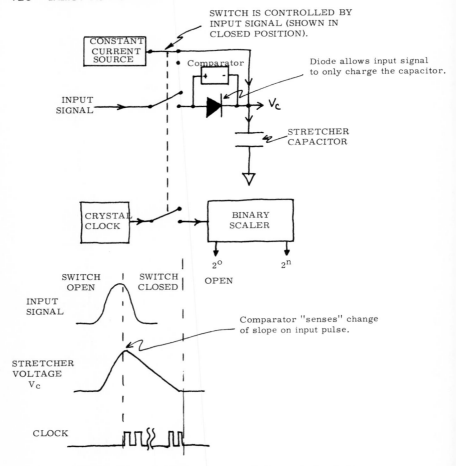

FIG. 3—*Height-to-time converter timing and signal diagram.*

With all the above logic conditions met, the "conditioned" input signal passes through the linear gate undistorted to the stretcher. The stretcher circuit then stores the peak voltage value of the input signal in the stretcher capacitor (as discussed in Fig. 3). The peak detector circuit monitors the input signal voltage and the stretcher capacitor voltage, and continuously compares the two signals until the input signal voltage drops below the voltage value "stored" on the stretcher capacitor. The change of signal levels between the two signals "triggers" the peak detector enabling the following functions: (1) the stretcher current source begins discharging the stretcher capacitor voltage to zero, (2) the address clock is connected to the scaler, and (3) the linear gate closes so no further input signals may enter the ADC. When the stretcher capacitor voltage discharges to zero volts, the address clock is "disabled," and a "store" command is generated to the memory system. As soon as the digital address scaler information is accepted by the memory, a "clear" command is returned to the ADC resetting all the circuits for the next input signal. For more detailed explanation of the different types of analog-to-digital converters available and a discussion of their circuitry the reader is referred to external texts.[2]

Specifications

In specifying an ADC for X-ray energy dispersive systems the user must understand the terminology used for the input signal requirements, the

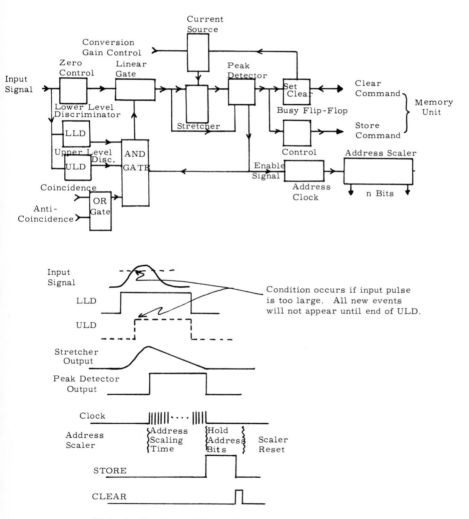

FIG. 4—*Typical height-to-time converter and timer.*

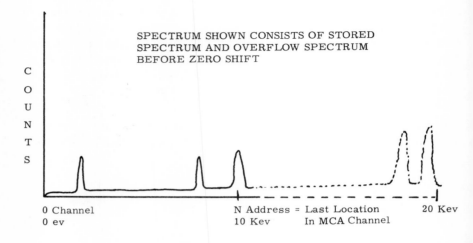

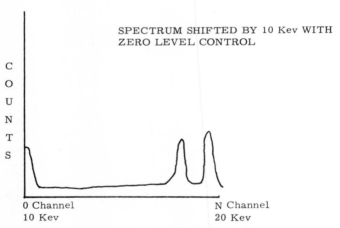

FIG. 5—*Zero level control used for spectrum energy shift.*

controls governing the converting of the input signal, and output capabilities to the memory unit.[3]

The normal acceptable input signal in most modern ADC's is an amplified pulse from the detector system. This pulse is specified by its amplitude, polarity, rise time, fall time, and output impedance from the source. The amplitude of the input signal usually is limited to 10 V due to the use of

[3] Heath, R. and Crouch, D. S., Idaho Nuclear Number 16923, Research and Development Report, Atomic Energy Commission, Nov. 1963.

solid-state electronics. The ADC channel conversion (resolution) is directly proportional to the amplitude of the input signal. ADC's presently are capable of resolving one part in 10,000; typically one part in 8192 resolution is possible for 8 to 10-V input. The polarity of the input pulse is usually positive to most modern ADC's due to the AEC[4] standardization. The "rise time" and "fall time" of the input pulse in the pulse amplitude to time ADC's is very flexible which contributes greatly to its popularity. Thus, a typical input pulse rise time can vary from 0.1 μs up to 10 μs with fall times in the range of 0.1 to 30 μs with little or no degradation to the resolution of the ADC. *Note*—the rise and fall time part of the input specification is very important and should be considered in great detail when looking at different ADC's for your detector system, since amplifier output rise and fall times will vary and could cause reduced resolution. The output impedance of the source (detector amplifier) should be as low as possible to limit cable noise pickup. Typical impedance values are in the range of 50 to 100 ohms which gives excellent termination for the ADC.

The coincidence input (available on some units) is used for "gating" the input signal on or off before entering the ADC. The signal level for the coincidence input is a d-c level and in most new ADC's conforms to the AEC standard of a positive 5-V signal with a time duration longer than the input signal. A typical application of the coincidence input would be as a monitor control for selecting specified energy events to the ADC.

The control functions include specifications on the lower level and upper level discriminator, the zero level, conversion gain, digital overflow, and on some units a baseline restorer control. These specifications are best explained by application requirements.

When setting limits on the input signals (energy range) to be accepted by the ADC, the discriminators will reduce the converting time normally spent on unwanted signals. Typically the range of the upper and lower level discriminators cover the amplitude range of the input signal acceptable to the ADC. In the newer ADC's the discriminators serve a secondary function as a single channel analyzer. The function of a single channel analyzer (SCA) is to set a lower and upper energy limit on the input events, thus allowing all events between these limits to be counted. The SCA becomes very useful and necessary when doing "line scans" or selective energy range studies on specimens in the electron microscope. Therefore, if the discriminators do serve as an SCA, the stability and sharpness of their cut-off limits is very important. Typical cut-off stability is 50 ppm, and the temperature stability is usually 200 ppm/C. In most ADC's the discriminator cut-off resolution is approximately one channel.

The zero level control, as previously discussed, should have at minimum the capability of adjusting the "zero level" (or zero channel energy intercept) from −1 to 100 percent of the energy input range acceptable to the

4 Report TID-20893, Atomic Energy Commission.

ADC. The temperature stability of the zero level is typically 50 ppm/C of full scale and 100 ppm/24h at constant temperature. In some ADC's there are digital zero offset capabilities. The digital zero offset has the same function as the zero level control except the digital offset is set by binary switch steps, usually with the coarsest step being half of the full scale range and dividing on down to one channel steps. The only drawback of using the digital zero offset is that a full conversion process is necessary in the ADC to establish the digital offset which increases conversion time and reduces counting efficiency of the system. The advantage of course is the precision of the offset steps.

The conversion gain control selects the magnitude of discharge current used for discharging the stretcher capacitor. Therefore, by changing the magnitude of the discharge current the input signal voltage on the stretcher capacitor can be amplified or attenuated by decreasing or increasing the discharge current, respectively, giving the effect of an external amplifier.

The digital overflow control is a safety precaution installed in the ADC to limit overscaling of the address scaler for too large an input signal for the conversion gain setting. If such a control is not available on the ADC the "overscaled" pulses will appear in the lower channels of the spectrum, thus giving distorted results. The overflow control usually is set to the size of the memory available thus guaranteeing compatibility between ADC and memory.

The baseline restorer in the ADC serves to increase input count-rate capability when input is a-c coupled and improve spectrum line definition by guaranteeing that the input signals recover quickly to a zero reference baseline. Most restorers are of the "passive" type which have recovery times of 50 μs to 0.05 percent of original baseline and 150 μs to 0.01 percent of the original baseline. If higher count rates are expected, an ADC should be purchased with an active baseline restorer. The active baseline restorer usually consists of an operational amplifier with "diode feedback" to the input. The restorer has typical recovery times of 3 μs to 0.1 percent of the baseline, therefore, improving spectrum definition at high count rates over passive restorers. If the baseline restorers are not incorporated in the ADC, the "pileup" or residue of prior pulses will be added to events currently being analyzed by the ADC, thus causing the reduced spectrum definition.

In defining the performance of the ADC, standards have been set describing the linearity of the system. These linearity standards are broken into two parts; integral and differential.

In defining integral linearity refer to Fig. 6 which depicts a plot of the ADC address number versus the signal input amplitude. An ideal ADC would have a plot as shown by the solid line, thus giving a perfectly linear ratio of address number to input signal amplitude. The solid plotted line represents a typical integral linearity plot. In defining the linearity of the

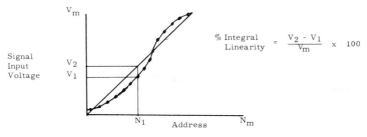

FIG. 6—*Integral linearity plot.*

ADC one is interested in the maximum deviation in any address from a straight line plot of address versus input pulse amplitude. Using this definition the integral linearity of Fig. 6 would be $V2$ minus $V1$ divided by V max, where V max is the maximum value of the signal input voltage. Typical integral linearities in a new ADC is from 0.01 to 0.05 percent over 99 percent of a 4000 to 8000 channel ADC.

Differential linearity describes the uniformity in address widths over the entire number of addresses in the ADC. The differential linearity usually is defined in percent of deviation of one address width to the average width of all addresses. Thus, for a differential linearity of 1 percent this means that the width of an address never departs by more than one percent from the average width of all addresses. Figure 7 represents a typical plot of address width deviation versus address number of an ADC with a definition of the limits to be expected. In most modern ADC's a differential linearity of 0.5 to 1 percent is typical for the top 99 percent of the addresses.

Outputs available from ADC's include a variety of signals. The most important signals available are the address bits defining the digital representation of the input signal and the store command indicating the event has been accepted and converted by the ADC. If the ADC includes a single channel analyzer, then a single output signal should be available with digital pulse levels of zero to +5 V for AEC compatibility. In some ADC's extra outputs are available to go directly to computers with the necessary interface. In this case the correct computer levels must be specified.

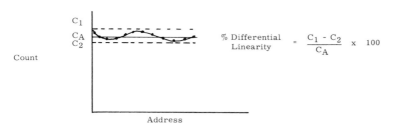

FIG. 7—*Differential linearity plot.*

Summary

In the discussion of the pulse amplitude to time converter for the ADC, please note that this is only one of several types of converters available on the market. It has been chosen for discussion because it has the most flexibility for X-ray energy dispersive systems. It has the capability of being the most accurate ADC available in the low price range. Its performance is high quality and ease in operation. With some of the accessories, such as a single channel output, coincidence gating, and computer output, the ADC becomes a very useful tool in a computer system and for "line scan" operations. For checkout procedures of the ADC, a fine article has been written by D. S. Crouch and R. L. Heath for routine testing and calibration procedures of multichannel pulse height analyzers.[3]

Memory

The memory is best defined as a high density data storage unit. Its main function is to sort and tally digital data presented to it from external sources such as an ADC, computer, time scalers, etc. The majority of the functions carried out in the memory unit are of digital form with operating times in the "micro" and "nanosecond" regions. Of course, for systems in the nanosecond region the system price is orders of magnitude higher than the microsecond systems. Typical memory systems of today operate with "memory cycle" times in the 1 to 5 μs ranges and are fully integrated circuit units.

In the development of the memory systems, two philosophies have evolved for storing information in the magnetic core memory. One is the "serial entry" memory which consists of serially loading a "multiple" of n digital bits into the memory to make up a total transfer of x bits of data to the memory. Example: $n = 4$, multiple $= 5$, therefore $x = 20$. The advantage of this system is its simplicity in memory drive circuitry and lower price. The disadvantage is slowness and incompatibility with parallel devices such as printers and computers. On the other hand, the "parallel entry" memory has faster data entry and exit times with external logic. The advantages of this system are its speed for transferring data from external units, such as the ADC, and its capability of presenting larger amounts of information to external display and readout units instantaneously. With the use of the integrated circuit and medium scaled integrates (MSI), the cost of the parallel entry memory is no longer a comparative factor. Therefore, in this paper the topic will be limited to the parallel entry memory since this unit seems to be the most popular in the X-ray energy dispersive and nuclear systems.[2]

Principle of Operation

Figure 8 is a block diagram for the parallel magnetic core memory and memory control circuitry. The system is broken into four parts consisting

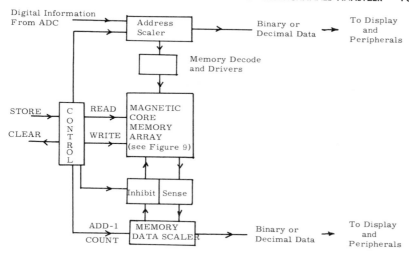

FIG. 8—*MCA parallel access memory control.*

of the magnetic core memory array, the address scaler circuitry, memory data scaler circuitry, and the control circuitry.

The magnetic core memory array is made up of a large number of cores arranged in a configuration to equal the number of addresses multiplied by word length (bits). For example, a 256 address analyzer with 20 bit count capacity per address would equal over 5000 cores or bits of data. For selecting and storing data in the magnetic core memory the system is orientated into a three axis configuration as shown in Fig. 9. The X and Y axis represent decoded address data. The Z axis represents the memory data counts. Thus for every X-Y line selected there will be a Z axis location selected. A coincident current "drive" on the X and Y axis is used to enter and remove data to and from the memory, respectively, as shown in the enlarged picture of a core in Fig. 9. Since the magnetic core exists in one of two magnetic states (either a "zero" or a "one" as termed by computer terminology), current must be directed through the core to magnetize it to switch from one state to the other. The characteristic of the magnetic core is changed from one state to the other by inducing a "switching" current, one half in the X and one half in the Y axis in coincidence, thus generating a total switching current I_s. This is called the memory read or write current I_r or I_w, respectively. If no information is to be written into the core, in other words writing a zero, an opposing current is introduced in the Y axis, called the inhibit current I_h, which will not allow the core to switch when the X and Y write currents are introduced. To "read" information from the core the current in the X and Y axes is reversed, and if the core had existed in a one state it will be switched to a zero. During the time the core is switching a small current is induced into a 4th wire through the core called the sense wire. The current in the "sense" wire is

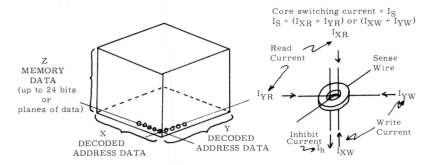

FIG. 9—*Magnetic core memory array.*

converted to a voltage and amplified, thus presenting a sense output to the appropriate bit in the memory data scaler. Thus by the proper selection of an X-Y line and presenting or accepting data to the Z axis, any core in the array can be selected and altered by this read/write operation. The selecting of these axes is controlled by the address scaler location, memory data scaler information, and the control circuitry of the memory unit.

As previously mentioned, the address scaler is the control for the X and Y co-ordinates driving the memory. The data supplied to the address scaler is either scaled-in serially or loaded in parallel from an ADC or an external device such as a computer. The address scaler also presents its information to a digital to analog converter for the X axis analog signal to the cathode ray tube (CRT) and plotter. Therefore, in selecting a memory unit the address scaler should have circuitry for "analog" decoding the digital information for display and be capable of accepting input data serially or in parallel. Resolution on the X axis analog information should be better than one part in five hundred if the address scaler is that large.

The memory data scaler in the memory unit plays an important information handling function since it "holds" the word information for each address location during display and data manipulations. The size of these words, depending on the memory system, can be from 12 to 24 bits, which is 4 thousand counts up to 16 million counts, respectively. The same information requirements for serial and parallel entry of the address scaler are also true for the memory data scaler. The memory data scaler information is supplied to a digital-to-analog converter for the vertical (Y) axis display on the CRT and plotter. The digital-to-analog decoders have typical resolution of at least one part in a thousand for display.

In most memory units that are not "computer compatible" the memory data scaler will be a binary-coded-decimal (BCD) scaler consisting of five to six decades. The reason for formatting the 20 to 24 bits of the memory data scaler into BCD is its availability to the outside world for "readout" to hard copy units, that is, printers, typewriters, etc. If the scaler information was presented externally in binary or octal format, converting would be necessary to get decimal numbers, and, therefore, increasing the memory price.

The control circuitry in the parallel memory functions as the programs normally used in the computer. However, the programs of the MCA are "hard wired" from front panel switches. The "switch" operations available in controlling the memory usually are broken into three program categories. (1) analysis, (2) readout, and (3) stop.

In the analysis mode, the memory is controlled from an external device such as an ADC, computer, or timer. In this mode the information is entered into the memory through the address and memory scalers. Since there are two analysis modes, pulse height analysis (PHA) and multichannel scaling (MCS), they will be discussed separately.

In PHA the memory unit operates with an ADC or multiple ADC's. When the ADC completes an event it generates a "store" command and the appropriate digital address. The memory, after receiving the store command, "sets" the ADC's address into the X-Y axis of the core memory and reads the Z axis information for that address from the memory to the memory data scaler. The new number in the memory scaler is now "written" back into the memory by the write command. This count adding process is continued until a satisfactory spectrum is accumulated from an X-ray energy dispersive detector system (Fig. 1).

The MCS differs from PHA mode in that each address location is sequentially selected as a function of time. The "time base" usually is generated from a crystal oscillator or external source. In MCS the data presented to the memory data scaler for each address are entered serially from an external source for a preset time set by the "time base." This mode can be used with the microprobe for giving "line scans" of a specimen. As shown in Fig. 10, the MCA and microprobe scan in synchronism. The information (or X-ray energy activity) from the SCA is scaled into the Z axis of the MCA giving a representation of the silicon concentration. *Note*—Energy window of ADC is set for X-ray energy of silicon.

The acceptable multichannel scaling input count rate varies according to the selected memory system and should be checked very carefully when purchasing such a system. It should be pointed out that in the serial machines, not being discussed, the input count rate limitation is usually 30 KHz or less. In the parallel units 10 MHz is typical, thus giving higher contour definition due to increase count rate per unit time. *Note*—The input count rate of the MCA is not presently the limiting factor in line scans. The amplifiers and detectors have shaping and recovery times that limit input rates to 10 to 20 KHz. However, with improved technology the higher count rates may be expected in the near future.

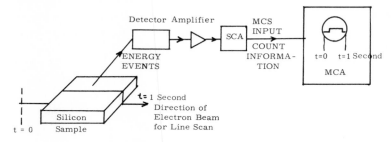

FIG. 10—*Line scan operation using MCA.*

In the *readout* mode of the MCA there are certain basic functions that should be included in every unit. These are the *display* mode, *pen* mode or *X-Y* recorder operation, *printout* (parallel printer capabilities), *type,* and *test.*

In the *display* mode the information in the core memory is interrogated sequentially and displayed on the CRT, thus an accumulated spectrum would be shown as in Fig. 1. The display for the multichannel analyzer is either built in as an integrated part or may be purchased separately as a stand-alone unit. Each display system has controls similar to those found on Tektronix or Hewlett-Packard scopes. In some of the more expensive MCA's, where isometric (3-dimensional) displays are possible, external display jacks should be available for connecting to TV screens or large cathode ray displays. Also, logarithmic converters are available for increasing the dynamic range of the vertical axis display information of the memory data scaler. Some log converters display six orders of magnitude of data at once. The vertical display should have a digital control allowing selection of different sections of the memory data register, thus giving discrete enlarged view of stored data. Although the display is a very simple part of the MCA, it is the easiest means of observing the memory data. The clarity of the display is invaluable in analyzing the data. The same displayed information should be available to the *X-Y* recorder (plotter) in the *pen* mode. The *print* mode will present sequentially the *Z* axis memory information from address zero to the last address in BCD form to the parallel printer. The *type* mode in the MCA operates the teletype or typewriter for hard copy readout. This mode requires that the MCA system have a parallel to serial converter built into its circuitry to make the data translation. The *test* mode is an interrogation of the core memory where a count is added sequentially to each address and memory data scaler to test the operation of the memory.

The *stop* mode in the MCA is a quiescent electronic condition guaranteeing that all logic functions in the unit are stopped. In this mode, transitions can be made to other operating states (*readout, analyze*) without affecting internal timing or loss of data.

Summary

The MCA of today is a versatile hard wired computer with an ultra-linear ADC as its transducer for the X-ray energy information from the electron microscope. When put to full use the MCA unit is capable of: (1) storing and displaying total X-ray energy spectra, (2) analyzing and recording elemental line scans on specimens in the electron microscope, (3) giving qualitative information on unknown elements in a specimen, and (4) storing 3-dimensional energy information from a specimen where multiple line scans are collected and assembled.

Of course, these listed uses are only limited examples of the applications now being done with the MCA in X-ray energy dispersive systems. As shown in the paper, the MCA is capable of being changed and controlled in a multitude of ways. Therefore, some of the features that should be included in an MCA to meet present and future requirements are listed here.

1. ADC should have resolution greater than present memory requirements. (Example, 8192 channel ADC and 1024 channel memory unit.)

2. ADC should have short conversion time for accepting maximum number of events from detector system.

3. ADC should have output flexibility so it may be used with memory or computer.

4. ADC controls should have control flexibility as described herein.

5. Memory read/write cycle time should be in range of ADC average conversion time to attain maximum system efficiency.

6. Memory unit should contain as many calculating features as possible, such as spectrum calibration, integration, spectrum comparison, etc.

7. For future expansion to computer the memory system should have full parallel input/output capabilities to its address and memory data scalers and control circuitry.

8. Memory units should operate with a variety of hard copy devices. Typewriters, punches, and tape readers are all commonly used devices.

In conclusion, the MCA manufacturers are striving to fill the X-ray spectrometer applications. This is being accomplished by the hard wired computer systems (MCA) and with expansion to the more versatile general computer systems.

Eric Lifshin[1]

Solid-State X-ray Detectors for Electron Microprobe Analysis

REFERENCE: Lifshin, Eric, **"Solid-State X-ray Detectors for Electron Microprobe Analysis,"** *Energy Dispersion X-ray Analysis; X-ray and Electron Probe Analysis, ASTM STP 485,* American Society for Testing and Materials, 1971, pp. 14–153.

ABSTRACT: Semiconducting X-ray detectors with low noise and high resolution have recently become available for the less than 20 keV range which is the region of interest encountered in the examination of X-rays produced in the electron microprobe analyzer. Although their energy resolution is generally less than that of crystal spectrometers, they offer special advantages including simultaneous collection of all detectable X-ray signals, high collection efficiencies, wavelength measurement independent of specimen position, and rapid collection of data in a form compatible with computer processing. Added to an electron microprobe, they provide a method of detecting in several minutes all elements with atomic number greater than sodium and present in quantities as small as a few tenths of a percent. Quantitative analyses are possible provided care is exercised in evaluating background spectra and peak overlap.

KEY WORDS: spectroscopy, solid state devices, semiconductors (materials), X-ray spectra, electron probes, background noise, quantitative analysis, X-ray analysis, scanning, electron microscopes, wavelengths, spectrometers, resolution, data storage, data processing

Semiconducting X-ray detectors have been used for high-energy gamma ray spectroscopy for more than ten years, but only during the last few years have detector systems with sufficiently low noise and high resolution been available for the less than 20 keV range, which is the region of interest encountered in the examination of X-rays produced in the electron microprobe analyzer (EMA) and the scanning electron microscope (SEM). The energy resolution of such detectors is still significantly inferior to that of standard focusing crystal spectrometers, but they have attracted widespread interest for the following reasons:

1. X-ray signals of all detectable wavelengths are collected simultaneously compared to crystal spectrometers which scan a spectrum, dwelling at any given wavelength for only a small fraction of the scan. This is im-

[1] Research staff, R and D Center, General Electric Co., Schenectady, N. Y., 12301.

140

portant because the minimum detectability limit depends on the total amount of X-rays collected as well as on the peak-to-background ratio of the particular spectral line being measured.

2. Higher counting rates than those obtained with a crystal spectrometer are possible because of greater detector efficiency, the elimination of intensity losses associated with crystal diffraction, and the ability to physically locate the detector closer to the specimen.

3. The detector responds to X-ray energy independent of specimen position. This eliminates the need to place a specimen on the focusing circle of a crystal spectrometer thereby making it easier both to examine irregular surfaces and to obtain more uniform X-ray scanning images free of defocusing effects.

4. Data can be collected rapidly and stored digitally in the memory of a multichannel analyzer in a form compatible with computer processing.

5. Improvements in detector crystals and preamplifier design are occurring so rapidly that energy resolution has improved from 600 eV a few years ago to less than 200 eV today. Thin detector windows or the complete elimination of detector windows can result in the detection of elements as low as oxygen in atomic number.

Combinations of solid-state X-ray detectors and various EMA's and SEM's have been reported by Fitzgerald et al [1],[2] Russ and Kabaya [2], and Solomon [3]. It is the purpose of this paper to describe methods of solid-state X-ray detector data collection, presentation, and evaluation when used on an electron microprobe and to examine some of the practical limitations encountered in spectral analysis.

Operating Principles

Electron Microprobe

In the electron microprobe a beam of electrons from an electron gun is focused by magnetic lenses to a fine spot, typically less than 1 μm, on the surface of a specimen. At the point of impact several processes take place including secondary electron emission, electron backscattering, and the generation of the characteristic X-ray spectrum. Qualitative chemical analyses involve the determination of which X-ray spectral lines are excited, while quantitative analyses require accurate measurment of line intensities. The exact position of the point being analyzed can be selected by a specially designed optical microprobe coaxial with the electron optical system. Images of the specimen also may be obtained in a scanning mode in which a deflection amplifier synchronously drives scanning coils in the electron optical column and the deflection coils in a cathode ray tube (CRT). The

---

[2] The italic numbers in brackets refer to the list of references appended to this paper.

brightness of the CRT can then be modulated by an electron signal (secondary, backscattered, or specimen current) or by a given wavelength X-ray signal to show surface structure or variations in chemical composition.

Crystal Spectrometers

Until recently the principal method of detecting X-rays was with a dispersive crystal spectrometer. A portion of the X-ray beam is allowed to pass from the specimen chamber through an exit port to the analyzing crystal where the various component wavelengths are diffracted according to Bragg's law:

$$n\lambda = 2d \sin \theta$$

where:

n = order of diffraction,
λ = X-ray wavelength,
d = interplanar spacing of the analyzing crystal, and
θ = diffraction angle.

Spectral scans are made by displaying X-ray intensity as a function of θ on a strip chart recorder. The values of θ corresponding to peaks then can be converted to wavelength or element from tables of Bragg's law calculated for many of the commonly used analyzing crystals such as LiF, quartz, KAP, PET, lead stearate, and so on. Because of the limited X-ray intensity generated in the specimen, microprobe spectrometers use crystal focusing optics. These optics require the use of a curved crystal which must be constrained with the specimen and X-ray detector to stay on the same circle called the focusing circle. To simplify quantitative analysis, an additional design requirement is imposed on the X-ray spectrometer system that the X-ray takeoff angle also be kept constant independent of θ. This increases the complexity of spectrometer design and has resulted in various ingenious schemes to mechanically control the motion of the analyzing crystal and detector. Such systems usually read directly in wavelength rather than θ.

The X-ray detector most commonly used in microprobes is the proportional counter. It consists of a gas filled tube with a wire running down the center held at a fixed positive voltage. As each X-ray photon enters the counter tube through a thin window, it ionizes the gas producing, on the average, n ions where

$$n = E/\epsilon$$

where:

E = energy of the X-ray, and
ϵ = average energy necessary to cause one ionization.

The electrons liberated are attracted to the central wire producing secondary ionization, and the total charge collected is converted to a voltage pulse which is further amplified. The amplified pulses then are converted to pulses of fixed size and time duration by a single channel analyzer so they can be counted by an X-ray scaler, displayed on a CRT, or transmitted to a ratemeter coupled to a strip chart recorder. It is important to emphasize the fact that with this type of spectrometer the energy selection is done principally by the analyzing crystal and that the function of the single channel analyzer is to eliminate higher order reflections and reduce background. Because X-ray energy is lost to processes other than gas ionization the energy resolution of the proportional counter is 15 percent at best.

Most microprobes are equipped with several crystal spectrometers each containing a selected crystal to cover a different wavelength range. In this manner X-rays from beryllium upwards may be detected. The minimum detectability limit varies from element to element and depends on the operating conditions of the instrument and the matrix composition. Under optimum conditions it is rarely less than 50 ppm and requires that the spectrometer be tuned to a given wavelength and the signal counted for 10 min or more. During qualitative analyses the spectrometers scan the spectrum over periods of 1/2 to several hours.

Solid-State X-ray Detectors

Detailed descriptions of the operating principles of solid-state detectors have been already covered in previous papers in this series and may be found in a number of sources [4,5]. The operation of the solid-state detector is similar to that of the proportional counter tube described previously, since the number of electrons produced is proportional to the incident X-ray energy. It should be stressed, however, that the applied bias potential does not cause secondary ionization and that the total charge collected is small compared to a proportional counter, and therefore requires a very sensitive low-noise preamplifier. Furthermore, pulse collection of all energies occurs simultaneously with pulse energies being separated electronically rather than by mechanical scanning, thereby placing stringent requirements on detector resolution and system linearity if it is to be useful in practical analyses.

Instrumentation

The system used to collect the data presented in this paper consisted of a CEC/Cameca microprobe, a Princeton Gamma-Tech detector, a Tennelec TC908 power supply, a Tennelec 202 BLR amplifier, and a Northern Model 630 multichannel analyzer equipped with modifications for both teletype and direct computer readout. The detector was inserted in the back port of the CEC/Cameca microprobe normally used for the con-

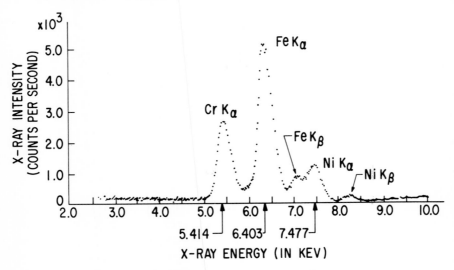

FIG. 1—*Spectra of Fe-20Cr-20Ni alloy.*

trolled oxygen leak, by means of a special adapter. A 1 mil beryllium window separated the specimen from the detector crystal. A typical result obtained with this system is presented in Fig. 1 which shows the energy spectrum of an Fe-20Cr-20Ni alloy as it appeared on the CRT display of the multichannel analyzer. Appropriate intensity and energy scales have been added.

Performance

As an initial measurement of performance, peak-to-background ratios and integrated line intensities were measured on the set of 20 standards supplied by the manufacturer. Data was collected at 20 kV and 5 nA beam current for 100 s. The choice of total counts contained within full width at half maximum (FWHM) was somewhat arbitrary based on a compromise between maximizing the total counts and minimizing the reduction in peak-to-background ratio. Recent work by Sutfin et al [6] indicates that choosing a window of 1.2 FWHM is optimum. The data obtained are shown in Table 1. Resolution in the system (which is about two years old) was measured at 315-eV FWHM and did not appear to vary significantly over the range of energies studied, implying that electronic noise was the primary factor limiting resolution. The sharp decrease in integrated peak intensity going from silicon to magnesium is due to absorption in the 1 mil beryllium. Using a 0.3 mil beryllium window not only improves these values but also makes it possible to detect sodium, neon, fluorine, and oxygen if used with a sufficiently low-noise detector.

TABLE 1—*Sensitivity of Li(Si) detector.*

ATOMIC NUMBER	X-RAY LINE	ENERGY (KEV)	PEAK TO BACKGROUND RATIO	*INTEGRATED PEAK INTENSITY (C. P.S. / NANOAMP)
12	Mg Kα	1.254	15.0	157.4
13	Al Kα	1.487	31.0	287.2
14	Si Kα	1.740	39.3	379.6
22	Ti Kα	4.510	27.7	279.4
23	V Kα	4.952	25.8	243.2
24	Cr Kα	5.414	28.6	237.9
25	Mn Kα	5.898	26.6	218.2
26	Fe Kα	6.403	27.4	188.3
27	Co Kα	6.930	23.5	175.9
28	Ni Kα	7.477	25.4	162.5
29	Cu Kα	8.047	22.2	141.5
30	Zn Kα	8.636	19.2	107.2
32	Ge Kα	9.885	16.6	89.2
40	Zr Lα	2.042	15.3	204.2
42	Mo Lα	2.293	16.3	200.4
47	As Lα	2.984	15.2	225.8
50	Sn Lα	3.444	14.7	236.6
74	W Lα	8.396	5.0	165.4
82	Pb Mα	2.347	6.9	317.3
90	Th Mα	2.997	7.1	351.7

* TOTAL COUNTS CONTAINED WITHIN FWHM

Several expressions for the minimum detectability limit C_{MDL} in microprobe analysis has been proposed. One of the most commonly used derived by Ziebold in [7]:

$$C_{MDL} = 3.29a/\sqrt{nTIR}$$

where:

n = number of measurements,
T = time period of each measurement,
I = net intensity,
R = peak-to-background ratio (peak intensity measured on a pure element standard and the background measured on the sample being studied), and
a = parameter in the Ziebold equation relating X-ray intensity to concentration (see Ref 8).

Since C_{MDL} depends both on total intensity and peak-to-background ratio, the high total intensity possible with the solid-state detector by simultaneously collecting pulses of all energies in many cases compensates for the low peak-to-background ratios compared to crystal spectrometers (39 to 1 versus 3000 to 1 on aluminum), thereby giving similar sensitivity during a qualitative scan. This is not surprising in view of the fact that during a typical crystal spectrometer scan at a uniform speed, less than 1/100 of the time is spent in traversing the peak of any given element. The counting rates encountered in practice were slightly better than those obtained with $10\bar{1}1$ quartz and KAP crystals, and a factor of five or six times better than those obtained with $10\bar{1}0$ and $11\bar{2}0$ quartz crystals. Placing the detector closer to the specimen without altering the construction of the microprobe could improve these rates by at least a factor of two. Since detector resolution can begin to degrade significantly for counting rates in excess of 10^4, the maximum permissible beam current is typically less than 10 nanoamperes (nA), low enough that spatial resolution is limited by beam penetration (1/2 to 5 μm), which depends on voltage and specimen composition, rather than beam diameter (less than 1/2 μm). This situation can be only partially alleviated by reducing beam voltage since voltages of at least two or three times the excitation potential are necessary to optimize peak-to-background ratio.

As a test of sensitivity, the spectra of eight steel standards of known concentration as listed in Table 2 were collected and analyzed. Measurements were made at 20 kV and 5 nA for 600 s. The electron beam was defocused and the specimen slowly scanned under it to eliminate the effect

TABLE 2—*Solid-state detector sensitivity for steel standards.*

STANDARD	Si	Mn	Ni	Cr	Mo	V	Cu
1	0.013	0.16	5.15	0.044	0.18	0.034	0.090
2	0.24	0.016	4.10	0.20	0.007	0.46	0.11
3	0.18	0.35	2.92	0.35	0.94	0.220	0.49
4	0.295	0.525	2.08	0.53	1.29	0.52	0.11
5	0.62	1.11	0.56	0.96	1.41	0.26	0.24
6	0.130	1.21	0.18	2.34	0.53	0.36	0.23
7	0.37	1.42	0.84	1.72	0.32	0.11	0.30
8	0.81	0.79	0.048	3.07	0.42	0.64	0.18

OPERATING CONDITIONS: 20 KV, 5na. BEAM CURRENT, 600 SECONDS.

☐ DETECTED WITHOUT SPECTRUM STRIPPING

⌐ ⌐ DETECTED WITH SPECTRUM STRIPPING

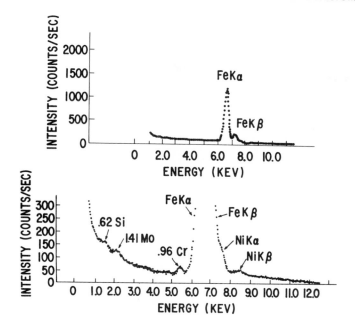

FIG. 2—*Energy spectra of steel standard 5.*

of in homogeneities. The spectra were transmitted to a computer where the data were smoothed by the method of Savitsky and Golay [9]. Peaks were detected using a method described by Barnes [10]. Table 2 shows the results. It can be seen that concentrations as small as 0.13 Si were detected. It is reasonable to expect that with higher resolution detectors, sensitivities of less than 0.1 percent should be possible under similar experimental conditions. Some difficulty in peak detection was encountered due to overlap as illustrated in the spectra of steel standard 5 given in Fig. 2. The upper part of the figure shows the spectrum as it appeared on the CRT display; only the iron $K\alpha$- and $K\beta$-lines are clearly visible. Additional peaks due to silicon $K\alpha$, molybdenum $L\alpha$, chromium $K\alpha$, nickel $K\alpha$ and nickel $K\beta$ can be seen with the scale expanded as shown in the lower part of the figure. The overlap between nickel $K\alpha$ and iron $K\beta$ is a serious practical problem. It could be overcome somewhat by stripping a spectrum of pure iron from the steel. This method revealed both the nickel $K\alpha$-peak obscured by the iron $K\beta$, and the manganese $K\alpha$-peak obscured by the iron $K\alpha$. Spectrum stripping produces a decrease in sensitivity for manganese compared to chromium, for example, due to statistical uncertainties in the iron signal accentuated by the subtraction process. Figure 3 further illustrates the problem of peak overlap with a series of synthetically constructed iron-nickel spectra obtained by adding standard spectra together to give the same 600-s total integration time at similar operating conditions.

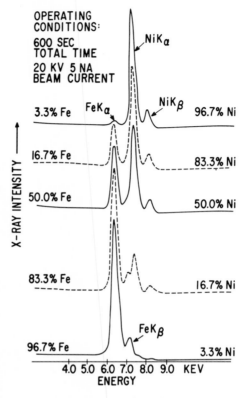

OPERATING
CONDITIONS:

600 SEC
TOTAL TIME

20 KV 5 NA
BEAM CURRENT

NiKα

3.3% Fe   FeKα   NiKβ   96.7% Ni

16.7% Fe   83.3% Ni

50.0% Fe   50.0% Ni

83.3% Fe   16.7% Ni

FeKβ

96.7% Fe   3.3% Ni

4.0 5.0 6.0 7.0 8.0 9.0 KEV
ENERGY

X-RAY INTENSITY ⟶

FIG. 3—*Combined Fe-Ni spectra.*

Consulting standard tables of X-ray energy [11] show that numerous peak overlaps of less than 100 eV (approaching the theoretical limit of detector resolution) can occur particularly when L and M series lines are present in addition to K-lines. To interpret such spectra the need for computer analysis becomes increasingly important. Figure 4 shows a computer separation of the spectrum of an Fe-30Cr alloy. Peaks were detected by the method described earlier, and then a least squares method was used to determine the Gaussian parameters of each peak described by functions of the form:

$$y = A_i \exp\left[-(x - M_i)^2/\sigma_i^2\right]$$

where:

 A_i = peak amplitude,
 M_i = peak mean,
 σ_i = standard deviation of the peak,
 i = refers to the ith peak,
 y = X-ray intensity, and
 x = energy or channel number.

The total spectrum can be described by:

$$y = \sum_{i=1}^{N} A_i \exp\left[-(x - M_i)^2/\alpha_i^2\right] + B(x)$$

where:

$B(x)$ = background.

Uncertainties in the shape of the background (which was fitted piecemeal) resulted in missing the chromium $K\beta$ which is only estimated in Fig. 4. Future refinements in the methods of spectrum analysis will hopefully overcome this type of difficulty.

To date, very few examples of quantitative analysis with solid-state X-ray detectors on microprobes have been published [12]. Table 3 contains the corrected relative intensities of both nickel $K\alpha$ and gold $L\alpha$ measured at 20 kV and an 18-deg takeoff angle both with crystal spectrometers and the solid-state X-ray detector. This system has no serious problems of spectral overlap, and the agreement between the two methods is good for specimens containing more than 6.3 weight percent nickel. Poorer agreement in the nickel data occurs for alloys with lower concentration due to the uncertainties in the background level which is comparable in size to the signal.

Data Presentation

Of particular value to the microprobe analyst are the numerous useful, and in most cases rapid, methods of data collection and presentation. The

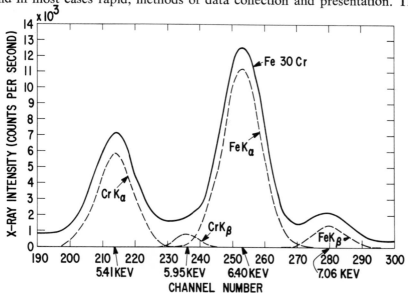

FIG. 4—*Computer separation of Fe-30Cr spectrum.*

TABLE 3—*Comparison between wavelength-dispersive and energy-dispersive X-ray spectrometers.*

SAMPLE	w/o Ni	w/o Au	CRYSTAL SPECTROMETERS		SOLID-STATE X-RAY DETECTOR	
			K_{Ni}	K_{Au}	K_{Ni}	K_{Au}
A1	87.7	12.1	0.8905	0.0915	0.8691	0.0925
A2	66.8	32.8	0.6482	0.2577	0.6418	0.2682
A3	47.6	52.9	0.4764	0.4355	0.4490	0.4584
A4	32.9	66.7	0.3159	0.5996	0.3197	0.6160
A5	14.7	85.3	0.1366	0.8084	0.1377	0.8093
A6	6.3	93.4	0.0572	0.9059	0.0814	0.8761
A7	1.1	98.7	0.0095	0.9925	0.0498	0.9775

system should first be calibrated by proper adjustment of the zero offset in the multichannel analyzer and amplifier gain to establish convenient operating conditions. A choice of 50 eV/channel, for example, will result in a spread of 25.6 kV in 512 channels of memory. A spectrum is then collected for a predetermined time period (clock time or live time) based on the desired degree of sensitivity. This spectrum then can be displayed on the CRT and the channel numbers corresponding to observed peaks read directly from the CRT, or with some analyzers read out directly, along with peak intensity on digital displays. Tables or graphs then may be used to identify the elements associated with the observed peaks. Since more than one spectrum may be stored in the analyzer memory, it is also possible to make direct comparisons between several spectra on the CRT display. A photograph with an oscilloscope camera of a specimen spectrum combined with appropriate standard spectra often serves as a useful record and convenient method of reporting results. More complete methods of data recording include the use of x-y magnetic tape recording, punched paper tapes, teletypes, high-speed printers, and direct computer collection. It already has been pointed out that many situations exist where further processing of the data is necessary. With more sophisticated interfacing, reference spectra or computer processed spectra may be returned to the memory of the analyzer where they can be visually displayed or readout by any of the methods previously described.

In addition to point analysis, a spectral line of interest may be chosen with a single channel analyzer (SCA) and its signal used to form X-ray distribution images or line scans. Figure 5 illustrates different methods of data presentation. For normal energy analysis the multichannel analyzer is used in the pulse height analysis mode (PHA), with the pulse sorted by

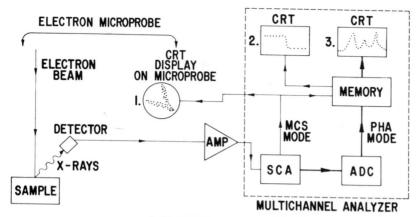

1. DISTRIBUTION IMAGE
2. LINE PROFILE
3. ENERGY SPECTRUM

FIG. 5—*Methods of X-ray data presentation.*

the analog to digital converter (ADC) and stored in memory. To obtain an X-ray distribution picture the window setting of the SCA is chosen to bracket the energy peak of the spectral line of interest. The output signal of the SCA, which consists of pulses of fixed size and time duration, then can be fed directly into the video amplifier of the EMA to produce X-ray distribution images on the CRT display, as shown in Fig. 6.

To obtain a line scan image, the electron beam is scanned across the specimen and the SCA output used to modulate the vertical deflection of the CRT as it scans a line, or the profile may be displayed on a strip chart recorder. An alternative method of recording a line scan is to use the multichannel scaling input mode of the analyzer. In this mode the beam is step scanned (or the specimen step scanned under a static beam), and the analyzer memory is used to record the X-ray intensity at each point. The resulting profile can be displayed on the analyzer CRT. This profile then can be photographically superimposed on a scanning electron image, as shown in Fig. 7.

Conclusions

The solid-state X-ray detector is a valuable addition to an electron microprobe. With it, rapid qualitative analysis for all elements with atomic number upwards of oxygen (depending on window thickness) can be performed. Sensitivity for many elements can be less than 1 percent for relatively short data collection times. If the results obtained in this manner are not adequate, sufficient insight into the specimen spectrum may be gained to reduce the time for subsequent crystal spectrometer examination.

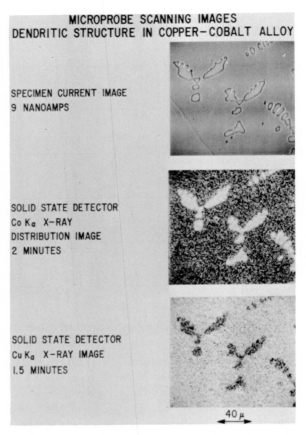

FIG. 6—*X-ray distribution images produced with the solid-state X-ray detector.*

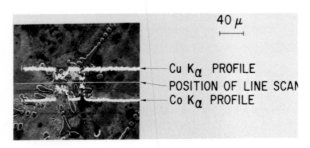

FIG. 7—*X-ray line scan image of Cu-Co alloy.*

Care must be exercised in interpreting spectra due to possible peak over-laps and uncertainties in characterizing background. Future improvements in detector resolution and refined methods of data collection and evaluation will result inevitably in still broader applications of this technique.

Acknowledgment

The author appreciates the assistance of M. F. Ciccarelli in collecting some of the data contained in this paper.

References

[1] Fitzgerald, R., Keil, K., and Heinrich, K., *Science,* Vol. 159, 1968, p. 528.

[2] Russ, J. and Kabaya, A., *Proceedings, Second Annual Scanning Electron Microscope Symposium,* Chicago, Ill., 29 April-1 May 1969, p. 59.

[3] Solomon, J. L., "Spatial Concentration Display and Mapping," *Fourth National Conference on Electron Microprobe Analysis,* Pasadena, Calif., 16-18 July 1969.

[4] *Proceeding of Eleventh Scintillation and Semiconductor Symposium,* Washington, D.C., *Transactions,* Institute of Electrical and Electronic Engineers, NS15, No. 3, June 1968.

[5] *Semiconductor Nuclear Particle Detector and Circuits,* Brown, W. L. et al, eds., Publication 1593, National Academy of Science, Washington, D.C., 1969.

[6] Sutfin, L. V., Ogilvie, R. E., and Harris, R. S., "Selection of Optimum Pulse Height Analysis Window," *Fourth National Conference on Electron Microprobe Analysis,* Pasadena, Calif., 16-18 July 1969.

[7] Ziebold, T. O., *Analytical Chemistry,* Vol. 39, 1967, p. 858.

[8] Ziebold, T. O. and Ogilvie, R. E., *Analytical Chemistry,* Vol. 36, 1964, p. 322.

[9] Savitzky, A. and Golay, M., *Analytical Chemistry,* Vol. 36, 1964, p. 1627.

[10] Barnes, V., *IEEE Transactions,* NS15, No. 3, 1968, p. 437.

[11] *Handbook of X-Rays,* Kaeble, E. F., ed., McGraw-Hill, N.Y., 1967.

[12] Myklebust, R. L. and Heinrich, K. F. J., "Qualitative and Semi-Quantitative Analysis with Non-Dispersive X-Ray Detectors," *Fourth National Conference on Electron Microprobe Analysis,* Pasadena, Calif., 16-18 July 1969.

J. C. Russ[1]

Energy Dispersion X-ray Analysis on the Scanning Electron Microscope

REFERENCE: Russ, J. C., **"Energy Dispersion X-ray Analysis on the Scanning Electron Microscope,"** *Energy Dispersion X-ray Analysis: X-ray and Electron Probe Analysis, ASTM STP 485,* American Society for Testing and Materials, 1971, pp. 154–179.

ABSTRACT: The scanning electron microscope is similar in many respects to the electron probe analyzer, but the scanning electron microscopist has a different goal and examines different specimens. The very low beam current of the scanning electron microscope requires the higher efficiency of the energy dispersion spectrometer (as compared to the wave-length dispersion type) resulting from a large solid angle, virtually 100 percent detection efficiency, and simultaneous analysis of all elements. Another advantage is that there is no confusion of lines due to the order of diffraction from a crystal. In examining rough surfaces or when working at low magnification, there is no loss of X-ray intensity because there is no geometric dependence of the energy dispersion detector. At high magnification the energy dispersion detector gives better X-ray resolution because of the smaller incident beam that can be used. The minimum detectable limits are comparable to wave-length spectrometers except for light elements, and quantitative analysis is practical. The modes of display include line scans and X-ray images in addition to point analyses. The paper discusses practical considerations in selecting and operating an energy dispersion X-ray system.

KEY WORDS: scanning, electron microscopes, electron probes, spectrometers, dispersing, X-ray analysis, X-ray spectra, resolution, electron beams, X-ray diffraction, quantitative analysis

The Scanning Electron Microscope and the Electron Probe Microanalyzer

The scanning electron microscope (SEM) is a relatively new instrument, available commercially for only about the last five years. In basic design (Fig. 1) it is similar to the electron probe microanalyzer (EPM); electromagnetic lenses are used to produce a fine beam of electrons and focus it on the specimen surface with a maximum beam energy of 30 or 50 kV. The beam is scanned in a raster pattern over the specimen in synchronization with a display cathode ray tube modulated with the signal

[1] Applications consultant, Jeolco (USA), Inc., Medford, Mass. 02155. Personal member ASTM.

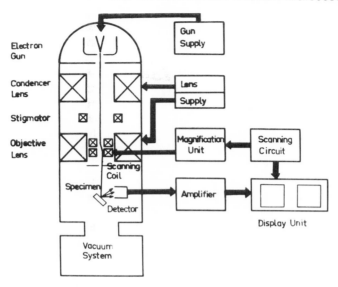

FIG. 1—*Block diagram of scanning electron microscope.*

from any of a variety of detectors for the various signals produced by the electron bombardment (Fig. 2).

The major difference between SEM and EPM is the intended use. The SEM is intended primarily as an imaging microscope, for the examination of surfaces at magnifications higher than possible with the unaided eye or the light microscope [1].[2] Since it is the surface topography that is of interest, the specimens are usually rough, and so the secondary electrons emitted from the surface are collected and amplified for imaging. The depth of field of the SEM is also very large (several hundred times that of the light microscope), and this is also important for examining rough surfaces. The rough specimens are held in a tiltable stage so that different viewing angles can be used.

By comparison the electron probe microanalyzer (as described in the preceding paper by Lifshin) is used primarily for analysis, and this emphasis requires the use of flat specimens mounted in a fixed geometry so that the necessary corrections can be made and the conventional type of X-ray spectrometer can be used.

The maximum magnification in the SEM is up to about ×50,000, much higher than in the microprobe. This requires a smaller incident beam diameter, generally under 100 Å in the SEM versus about 1 μm in the EPM. The reduction in beam diameter is accompanied by a reduction in

---

[2] The italic numbers in brackets refer to the list of references appended to this paper.

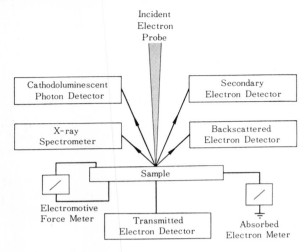

FIG. 2—*Signals generated by impingement of electron beam on a specimen.*

beam current, since for a given gun brightness and lens design the diameter is proportional to the current to the three-eighths power:

$$d = (2C^{1/4}/\pi^{3/4}) (i/B)^{3/8}$$

where:

C = spherical aberration constant or electromagnetic lens,
i = beam current,
B = gun brightness = $S(eV/kT)$,
V = accelerating voltage,
T = temperature of source, and
S = electron emissivity of source.

Hence the beam diameters of the EPM vary from 1/2 to several μm for beam currents of 10^{-6} to 10^{-8} A. To achieve beam diameters of 100 Å or less the SEM beam current is 10^{-11} to 10^{-12} A. This reduction of current causes a proportionate reduction in all of the signals generated including the characteristic X-rays, and this point will be discussed in more detail.

The differenec in primary function of the SEM and EPM usually is accompanied by very different outlooks on the part of the operators. The EPM user is inclined analytically and is often familair with other complicated analytical techniques. Many, though by no means all, SEM users are in other fields such as biology, industrial quality control, etc., and have little other contact with X-ray analytical instrumentation. Therefore, they require that the operation be as simplified as possible, short of sacrifice in performance.

Energy Dispersion Versus Wavelength Dispersion Spectrometers

Efficiency

Geometric—Because of the differences between the SEM and EPM, the application of energy dispersion (ED) X-ray spectrometer to the SEM is not exactly the same as to the EPM, and the comparison of ED versus wavelength dispersion (WD) X-ray spectrometers emphasizes different features of each from those discussed in the preceding paper by Lifshin.

First and foremost, the ED spectrometer is better suited to the low X-ray fluxes encountered in the SEM. Because of the exacting geometric requirements of Bragg diffraction, the WD spectrometer uses collimators that restrict the analyzed X-ray beam to a very small solid angle. Even for ideal Johannsen X-ray focusing conditions in the WD spectrometer the solid angle rarely can be made larger than 0.0001 steradian. This type spectrometer gives X-ray count rates of up to about 10^5/s for a 10^{-7}-A beam on a pure element. At the beam currents of the SEM this would be a count rate of about 1/s, far too low for useful work.

The ED spectrometer, on the other hand, is not limited in any way by geometry. The only collimation required is a shield to prevent X-rays excited by backscattered electrons that strike other parts of the specimen chamber from reaching the detector. Hence the detector can be positioned very close to the specimen to cover a large solid angle (up to 0.01 steradian). The resulting count rates are thus increased by a factor of about 100 times by this means alone.

Detection—Furthermore, the detection efficiency of the ED detector is much higher than the WD. The proportional counter efficiency varies with energy, and the diffraction of X-rays from a crystal is inherently inefficient (typically less than 20 percent). The result is a loss of intensity and a variation through the spectrum that can complicate analysis unnecessarily. By comparison the ED detector is virtually 100 percent efficient over a considerable energy range. The drop at high energies is not of concern in SEM (or EPM) use because these instruments operate only up to 30 or 50 kV and efficiently excite X-rays only up to about half that energy. The drop in efficiency at low energies is caused by absorption in the beryllium window and will be discussed further below.

The combination of higher geometrical and detector efficiency means that ED detectors can give count rates for a pure element up to several hundred times greater than the WD spectrometer. This makes it practical to use the ED X-ray analyzer on the SEM without increasing the beam current from the 10^{-11} to 10^{-12} A range normally used for high resolution microscopy (the low current also minimizes specimen damage and contamination).

Spectral Analysis—The efficiency of the ED spectrometer is further increased by its simultaneous processing of all X-rays from all the elements present. When the WD spectrometer is used to analyze an unknown it must be traversed through a range of 2θ values, and this must be repeated with several crystals (typically 3) to cover the elements from sodium to uranium. This is time consuming (taking from one half to several hours) and means that only a small fraction of the total analyzing time is spent on each element. The ED spectrometer can accumulate counts for each of the several elements present in the same time period, and a 100-s counting time with typical SEM beam currents will permit detection and identification of the major and minor elements present. Longer times up to about 15 min can be used to detect trace amounts.

In the EPM the high potential efficiency of the ED detector resulting from large solid angle, high detection efficiency, and simultaneous handling of all X-rays cannot be realized. The primary design criterion for the EPM is the generation of the greatest possible X-ray flux, and with beam currents of 10^{-6} A the count rate for a pure element measured with the WD spectrometer is high enough to create significant dead time in the counting electronics. This problem is compounded in the ED system, and with beam currents above about 10^{-8} A the dead time can become so severe that it completely limits the system, so that the beam current in the EPM must be reduced when the ED detector is used. However, in the SEM the beam currents are low; far from overloading the electronics, the problem has been one of obtaining high enough count rates for practical X-ray detection. The full potential of the ED spectrometer thus can be used with the SEM and yield rapid and satisfactory analysis.

Light Element Detection

The principal limitation of the ED spectrometer is its ability to detect light elements. The spectral response of the detector, discussed previously, shows a sharp dropoff at about 1 keV. With conventional 1 mil (0.025 mm) thick windows of beryllium only about 15 percent of the incident sodium X-rays (1.04 keV) enter the detector. (The dead layer of the detector also contributes some absorption but much less than the window.) With this thick a window the detection of X-rays from elements lighter than fluorine is very difficult. Oxygen (0.52 keV) has been detected in laboratory experiments [2], but the count rate and detectability were poor. It is possible, of course, to use thinner windows. Some experimentation with 1/3 and 1/2 mil (0.008 and 0.012 mm) beryllium and stretched films of nitrocellulose and polypropylene has been performed, but problems with porosity and lack of strength have not yet been entirely overcome.

It is conceivable that the window might be eliminated altogether. The SEM operates in a vacuum, and the window is not necessary if suitable airlocks (either on the ED spectrometer or the SEM) are provided for specimen introduction. The vacuum in most SEMs contains oil and water vapor, but suitable cold surfaces could be placed near the detector to prevent contamination from building up on the Si(Li) diode. The high energy backscattered electrons from the specimen could be deflected from striking the detector (which they would ionize) by a suitable electrostatic or electromagnetic field (this would have to be arranged so as not to affect the electron optics): and the problem of fluoresced visible light from the specimen (which would also excite the detector) may not be objectionable for most specimens because its intensity is very low.

All of these design problems undoubtedly could be overcome, but other more basic limitations remain. The first is the low energy noise of the ED system, a peak centered at zero energy and extending up to 2 or 3 times the full-width-at-half-maximum (FWHM) resolution of the system. Even for a 100-eV system (well beyond the present state of the art) this would make it impossible to resolve carbon X-rays (282 eV) from the noise peak.

Furthermore, these light elements (carbon, nitrogen, oxygen) are often present in a matrix of heavier elements, either metals or minerals. The L-lines of the elements from atomic number 16 to 30 are all in the range from 160 to 1000 eV and so are the M-lines from heavier elements. Hence, it may be impossible to resolve the light element peaks from these other peaks. For these reasons it may be desirable to use other means of detection for light elements, either flow proportional counters, as described in the paper by Sutfin and Ogilvie, or WD spectrometers.

Rough Surfaces

The geometric X-ray focusing requirements of the WD spectrometer restrict most EPM specimens to flat polished surfaces. This also makes possible the calculation of absorption and fluorescence corrections to the X-ray data for quantitative analysis. Polished specimens are sometimes examined in the SEM, and there is some call for using the SEM as a microprobe analyzer to perform quantitative analysis, but the main thrust of SEM work is toward rough specimens. It is difficult to use WD spectrometers on rough surfaces because the X-ray source must be on the Rowland circle of the spectrometer. This requires vertically shifting the specimen to bring the feature of interest into focus and for a rough surface would make WD analysis prohibitively cumbersome. Figure 3 shows that a vertical displacement of only 40 μm can cause defocusing of the X-ray optics and change the count rate by 50 percent [3].

By comparison the ED spectrometer, as mentioned before, has no such geometric requirements; so that vertical displacements or specimen relief cause no important change in count rate. The only effects are the slight change in takeoff angle and thus in absorption of the X-rays. Since rough surfaces do not lend themselves to true quantitative analysis but to qualitative or semiquantitative analysis, these effects are less important.

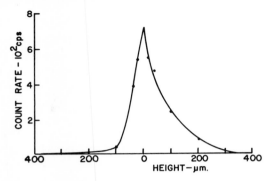

FIG. 3—*Variation in WD X-ray count rate with vertical displacement of specimen.*

Takeoff Angle

Specimen relief can cause further problems if the X-ray takeoff angle is low, because some areas will not be visible to the detector. This can be alleviated somewhat by tilting the specimen, which is possible in most SEMs. However, this may limit the ability of the instrument to view the surface and for very rough specimens still is not entirely adequate. The answer is to position the detector as high as possible over the specimen. High takeoff angles are achieved in some electron probe microanalyzers at the cost of design complexity; but the SEM is designed primarily for purposes other than X-ray analysis, and high takeoff angles for WD detectors generally are not provided. The ED detector can be designed more flexibly than the WD, and it is possible to place the detector much higher over the specimen, limited only by the ED manufacturer's ingenuity (X-ray takeoff angles of 45 deg from a horizontal specimen are available for some ED-SEM combinations).

Low Magnification

Most electron probe microanalyzers have a minimum magnification of about ×400 to ×500 at which image displays of X-ray analysis are possible. This is because the focusing requirements are quite restrictive and parts of the specimen that deviate from the Rowland circle give reduced X-ray intensity. Figure 4 shows an aluminum specimen machined with 0.0005-in. steps [4]. The WD X-ray image shows an apparent change in concentration that appears to relate to one of the steps but is actually due to part of the area being out of focus in the WD spectrometer. The ED X-ray image shows correctly a uniform concentration. The width of the shadows is dependent on takeoff angle.

Figure 5 shows another example. This polished mineral section (ilmenite) was examined to determine the percentage of grains rich in titanium. Even at the lowest EPM magnification only a few grains could be seen at a time, and a prohibitively large number of fields would have to be ex-

FIG. 4—*Images of machined aluminum block showing drop in intensity for areas not in focus in WD spectrometer: (a) secondary electron image, (b) WD X-ray imag, (c) ED X-ray image.*

amined. Since the SEM is intended by its very nature to operate down to much lower magnifications and since the ED analyzer is not geometry sensitive, the SEM-ED combination is well suited to this type of specimen. At ×100 only 1/25 as many fields need be examined as at ×500, with less need for accurate stage stepping and a great time savings.

High Magnification

When the SEM is used in its usual mode with secondary electron imaging, the image resolution is 100 to 200 Å. However, the X-ray emitting

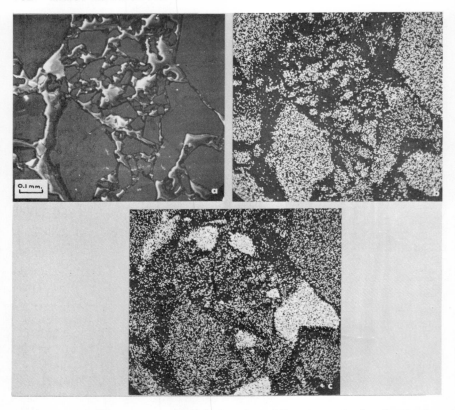

FIG. 5—*ED X-ray images of polished section of ilmenite, a titanium-bearing iron ore:* (a) *secondary electron image,* (b) *iron Kα X-ray image,* (c) *titanium Kα X-ray image.*

volume is much larger than that because the X-rays come from nearly the entire electron capture volume. Scattering of the incident electrons in the specimen makes the diameter of this volume larger than the incident beam. At large beam diameters the capture volume varies with beam diameter; but as the incident beam becomes very small the capture volume is limited by the incident energy of the electrons, which controls the distance the electrons can scatter in the material.

Figure 6 shows that the X-ray emitting volume for typical SEM beam diameters is limited by accelerating voltage and is about one quarter the size of that typical of the EPM [5]. This means that X-ray analysis can be obtained in the SEM from smaller areas. Figure 7 shows an example of this: the particles are 3000 to 4000-Å chromium-rich carbides that nucleated the dimple rupture fracture. The particle analysis shows no nickel, which is present in the stainless steel matrix, showing that the analyzed volume is smaller than the particles.

Single Element Monitoring

The ED detector also can be used to monitor the count rate from a single element using a single channel analyzer (the SCA may be either a separate instrument or a part of the multichannel analyzer). This output can be used to modulate the display cathode ray tube to show the same kind of X-ray distribution image commonly used for the EPM, as illustrated in Fig. 8. The high efficiency of the detector gives adequate count rates for this kind of imaging at typical SEM beam current. Line scans are also possible (Fig. 9). Used with a ratemeter or integrating ratemeter the output can be also used to represent the element concentration on a continuous linear scale and is suitable for modal analysis.

Modal analysis (also known as quantitative metallography or stereography) is the technique of determining the structural parameters of a solid, such as grain size and volume fraction of various constituents, average contact area of spacing between phases, etc., [6]. Measurements usually are made on polished specimens, and the method and instrumentation are identical for the SEM and EPM. The advantage of the ED spectrometer is that it gives higher count rates and can be used at both higher and lower magnifications, whichever is appropriate to the specimen.

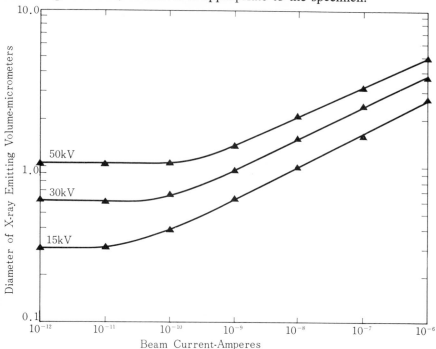

FIG. 6—*Spatial resolution of X-ray signal for iron Kα X-rays from steel specimen, showing effect of beam current (beam diameter) and accelerating voltage.*

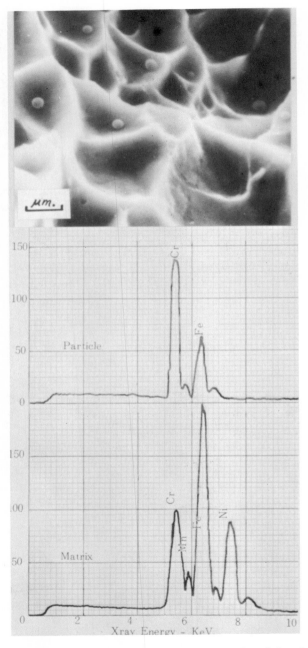

FIG. 7—*Dimple rupture fracture in stainless steel nucleated by chrominum-rich carbides.*

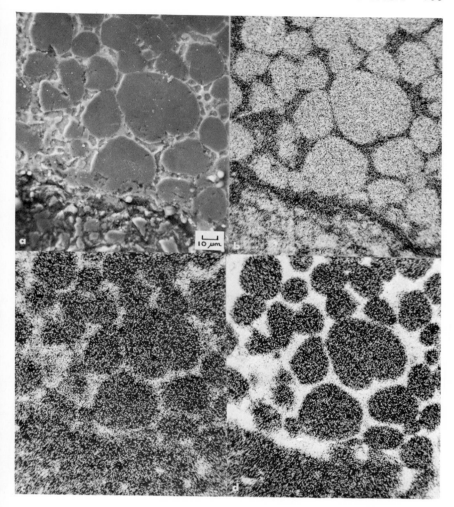

FIG. 8—*Refractory brick composed of magnesium oxide particles in calcium sili-cate binder:* (a) *secondary electron image,* (b) *magnesium* Kα *X-ray image,* (c) *silicon* Kα *X-ray image,* (d) *calcium* Kα *X-ray image.*

Monitoring of any number of elements desired (using a single channel analyzer and ratemeter for each element) is easy with the ED system, and when desired the output can be plotted on a strip chart recorder or other-wise recorded.

Trace Element Analysis

One of the particularly important criteria for any analytical system is its ability to detect trace amounts of elements present. In comparison to the WD spectrometer, the ED spectrometer produces spectra that have poorer

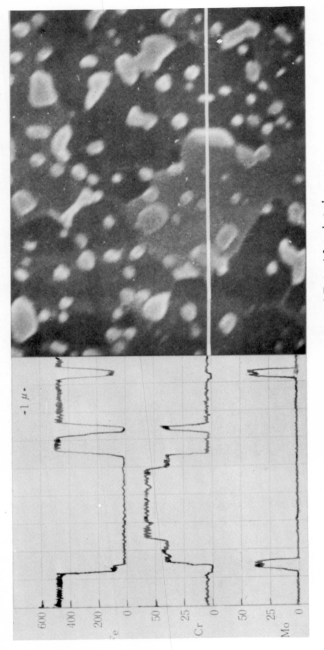

FIG. 9—Line scan analysis of Type A-2 tool steel.

peak-to-background ratios and higher count rates, and it is important to consider the effect of these factors on detectability.

A useful figure of merit can be obtained by expressing the fractional standard deviation of the X-ray count for any one element as:

$$\sigma = \sqrt{\frac{I_P + I_B}{(I_P - I_B)^2}} \sqrt{\frac{1}{t}}$$

where:

I_P = peak count rate,
I_B = background count rate, and
t = counting time.

Then if $I_o = I_P - I_B$ is the net count rate, we can reduce this to

$$\sigma = \sqrt{\frac{1}{t}} \sqrt{\frac{1}{I_o} + \frac{2I_B}{I_o^2}}$$

When I_o is small compared to I_B and we approach the minimum detectable limit, the latter term predominates, and σ will be the smallest (and our ability to detect small amounts of the element best) when I_B/I_o^2 is as small as possible. Inverting the expression, we see that I_o^2/I_B, which now we want to maximize, is the product of I_o, the net count rate, and I_o/I_B the net peak-to-background ratio. This product is then the figure of merit for comparing ED and WD spectrometers.

Table 1 lists some representative data for WD and ED spectrometers. The WD data were taken from published catalogs of various EPM manufacturers. The ED data were measured with a system (Nuclear Diodes XS-25-JSM-U3) installed on an SEM (Jeolco JSM-U3). The results show that the figure of merit for the ED detector is as good as or better than WD, so that minimum detectable limits are better. This is especially true for the heavier elements, which require in the WD spectrometer the use of higher order diffraction lines with considerable loss of

TABLE 1—*Comparison of WD and ED spectrometers.*

	WD Spectrometer[a]			ED Spectrometer		
Element	Count Rate, cps/pA	P/B Ratio	I_o^2/I_B	Count Rate, cps/pA	P/B Ratio	I_o^2/I_B
Al	0.7	1200	840	7.6	160	1216
Ti	1.1	1500	1650	16.6	135	2241
Cu	0.25	400	100	11.6	104	1206
Mo	0.03	40	1.2	1.22	47	57

[a] Catalog values for count rate and P/B generally are measured with different operating conditions (slit, PHA window, etc.) and cannot both be optimized at once.

intensity. The ED spectrometer figure of merit typically exceeds the WD by 50 to 100 times for heavier elements.

This advantage is limited in the EPM because the high X-ray flux generated and the need for processing the entire spectrum at once produce dead time in counting electronics and reduce the effective count rate. However, in the SEM at its normal operating beam currents this is not the case.

Furthermore, the EPM excites a volume of several cubic microns weighing about 10^{-11} g. In the SEM the exicted volume is, as shown above, about one third the diameter and of the same depth so that it weighs about an order of magnitude less, or 10^{-12} g. Since for most elements the detectable limit with ED is 0.1 percent or less, this means that as little as 10^{-15} g can be detected. No other analytical technique has the absolute sensitivity afforded by the SEM-ED combination.

Biological Specimens

One particular area of application in which this detectability is of interest is for biological specimens. The presence of trace amounts of compounds containing metal ions in various intracellular structures has been suspected for some years. Elaborate chemical staining or autoradiographic techniques in some cases can detect them, but not with adequate sensitivity or spatial resolution. Electron probe microanalysis has been used in some cases but also lacks adequate spatial resolution; and, furthermore, the high beam current can burn up delicate tissue specimens.

Figure 10 shows a typical application of ED analysis to biological tissue in the SEM.[3] Proof that the capillary ran along the bone/cartilage interface was obtained by identifying the bone and cartilage with the ED spectrometer. Also, besides looking at bulk specimens, the SEM can be fitted with an attachment to produce transmitted electron images similar in many respects to the conventional transmission electron microscope. There are some differences in favor of one or the other technique, but they are not of concern here. The resolution is on the order of 100 Å, adequate to identify the structures of interest. Better image resolution would not be helpful because the X-ray spatial resolution is not that good.

Figure 11 shows an example of his technique [8]. Using a thin section of tissue (from a few hundred to a few thousand angstroms thick), the electrons do not scatter as far laterally as in a bulk specimen, so that the excited volume is smaller in diameter. The spatial resolution of X-ray analysis is approximately equal to the thickness of the section used. Thinner sections give better resolution but contain less material so that longer counting times must be used. (This technique is not limited, of course, to

[3] Courtesy of Drs. I. Redler and M. Zimmy, Louisiana State University Medical School.

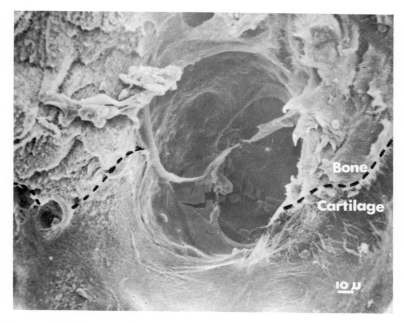

FIG. 10—*Capillary (containing red blood cells) between bone* (top) *and cartilage* (bottom). *Dashed line shows interface, identified by ED X-ray analysis of calcium in bone and sulfur in cartilage.*

biological specimens. It also can be used with thin metal foils or replicas containing particles.)

The ability to detect trace amounts is even further improved in this case because there is little absorption of X-rays in the tissue and the background radiation is low from the low atomic number material. This, combined with the even smaller volume excited, makes possible the detection of less than 10^{-16} g of an element.

Quantitative Analysis

For performing quantitative analysis the criterion for the X-ray spectrometer is again that σ be small, but now I_o is large so that the $1/I_o$ term dominates $2I_B/I_o{}^2$. Hence, we want a high total count rate, and, as explained before, the ED spectrometer excels in this regard. The important point to be made is that quantitative analysis is performed in just the same way for the ED and WD data. The corrections for fluorescence, absorption, and atomic number effect are required for both. Corrections for detection efficiency, etc. are different but necessary in each case.

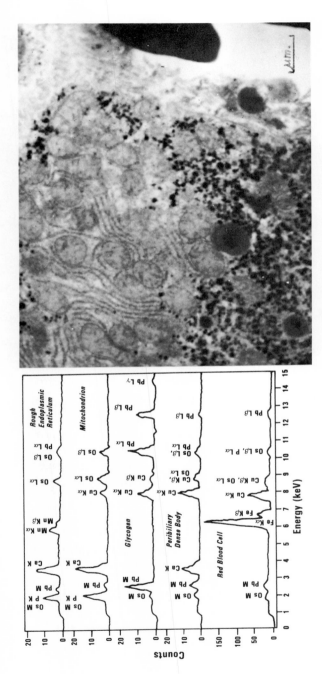

FIG. 11—*Transmission SEM image of cirrhotic rat liver with ED X-ray analysis of intracellular structures.*

One possible difficulty is the greater likelihood of interfering peaks with the ED spectrometer, since the resolution is poorer. This is generally not a serious limitation in identifying peaks and often can be overcome for quantitative work in those cases where interferences do occur.

One technique is to separate the overlapping peaks using deconvolution techniques like those described in the paper in this volume by Heinrich and Myklebust. It is also possible to use the multichannel analyzer of the ED spectrometer itself. An electronic pulser is used to simulate X-ray pulses and produce an artificial spectrum of peaks. It is adjusted to produce peaks of the same spread as the detector, and by changing the setting of the pulser output the peak can be made to appear at any position in the spectrum. Then the pulser (spectrum reconstruction generator) is used to build in one memory segment of the analyzer, a spectrum composed of known peaks to match an unknown overlapped spectrum stored in another memory segment.

A far faster technique that can be used in some instances is to calculate the size of one of the interfering peaks from another peak from the same element that has no interference and subtract it from the total count of the composite peak. This approach is limited, of course, to situations where all but one of the interfering peaks belong to elements for which another peak without interference is present. Also, the ratio of the two peaks must be well known, and this can depend on the accelerating voltage used and the relative absorption of the peaks by the other elements in the specimen.

An example of the satisfactory use of this technique is the analysis of brass specimens by ED spectrometry which required subtracting the copper $K\beta$ from the zinc $K\alpha$ peak to obtain the correct zinc count rate [9]. This was done by extending the range of integration over the total composite copper $K\beta$ + zinc $K\alpha$-peak and then subtracting 14.4 percent of the net copper $K\alpha$ count. This copper $K\beta$: $K\alpha$ ratio was measured under the same operating conditions (detector configuration and takeoff angle, and accelerating voltage) on a pure copper specimen; and it was used because the relative absorption of the two lines was much the same in the brass specimen as in the pure copper.

The net integrated peak intensities for copper, zinc, and the minor elements, aluminum, iron, tin, and lead, then were used in an iterative correction program based on linear equations and empirical coefficients. Table 2 shows the matrix of correction coefficients determined with sintered powder specimens. The ED analyses gave results with 72 percent less than 1.5 percent relative error for the major elements and 69 percent less than 0.4 percent absolute error for iron and tin, and 60 percent less than 0.7 percent absolute error for aluminum and lead. These results are comparable to those for WD quantitative analysis but required only 100 s of analyzing time (for all elements simultaneously), far less than the WD method.

TABLE 2—*Correction coefficients for brass specimens.*[a]

Matrix Element	Cu	Zn	Fe	Sn	Al	Pn
Cu............	0	1.22	1.06	1.81	1.03	1.99
Zn............	1.16	0	1.05	1.80	1.03	2.00
Fe............	0.60	0.63	0	0.97	1.02	1.65
Sn............	1.22	1.24	1.07	0	1.04	2.09
Al............	4.05	4.07	3.75	5.36	0	8.3
Pb............	0.97	0.98	0.95	0.92	1.15	0

$$^a\frac{C}{K} = a + C\,(1 - a) \qquad a_{\text{effective}} = \sum_{i}^{n} a_{ki}\,\frac{C}{(1 - C_k)}$$

where:

C = elemental concentration,
K = ratio of measured X-ray intensity to intensity from pure specimen,
a = correction coefficient, shown in table.

Ease of Operation

Because the SEM operator generally has a different background and purpose from the EPM user, it is important to consider the ease of operation of the ED system. First and foremost, the ED system presents a very simple, uncluttered spectrum. There are only the $K\alpha$, $K\beta$, $L\alpha$, $L\beta$, $L\gamma$, and M-lines to identify, and this is usually easy because of the appearance of the K's as a pair, the L's as a triplet, and the M as a broad peak. In the WD spectrometer, on the other hand, diffraction can occur at any angle for which the Bragg equation $n\lambda = 2d \sin \theta$ is satisfied. Since "n" can have any integer value, many "orders" of diffracted lines are possible (up to $n = 8$ or more can be important for some crystals). This results in a very cluttered spectrum with many potential interferences between different order lines from different elements, and it is often necessary to check more than one spectrometer setting to unequivocally identify the element present.

Furthermore, the cluttering of the spectrum with many lines requires the use of thick comprehensive tables to identify angles with elements [10]. The ED spectrum is much simpler in appearance. The K, L, and M energies increase smoothly and monotonically with atomic number, as shown in Fig. 12, and can be easily summarized in a one page chart. Absolute systems of calibration in which particular channels in the analyzer always correspond to the same energy (for example, 20, 50, or 100 eV per channel) make it easy to identify peaks quickly, and most operators learn to identify the lines of many elements on sight. When more than one line from an element is present, it is usually easy to identify, as shown in Fig. 13.

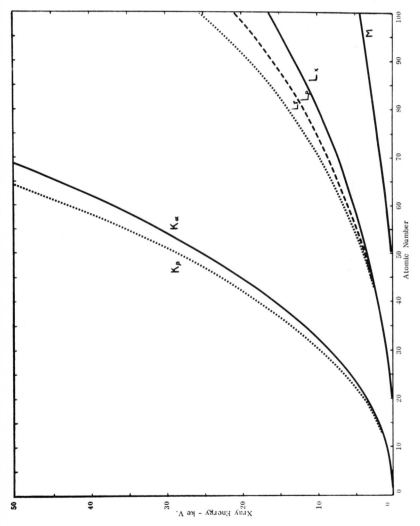

FIG. 12—*Variation of X-ray energy with atomic number.*

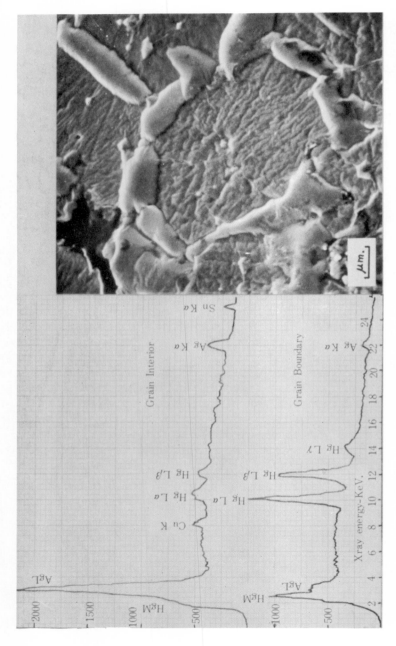

FIG. 13—ED X-ray analysis of dental amalgam. Pressure application caused segregation of mercury to grain boundaries.

This subjective rather than technical advantage of ED over WD X-ray analysis is important particularly for application to the SEM, since X-ray analysis is an auxiliary technique and should not demand too large a portion of the operator's attention and training. Along the same lines, the ED system does not require the elaborate precision hardware of the WD spectrometer and thus eliminates the need for alignment of crystals and other maintenance. It is possible to design the ED system operation controls to be few in number and easy to understand without compromising performance, further enhancing its ease of operation.

Parameters for ED Operation and Selection

Accelerating Voltage

To obtain the best results with an ED X-ray spectrometer on the SEM it is necessary to use the proper instrument operating conditions. Of these the most important is the accelerating voltage. The energy of the incident electrons, of course, must be high enough to excite the X-rays from the element of interest, but beyond that there are other factors to be considered.

For specimens in which the best possible spatial resolution is desired, the lowest accelerating voltage that will give adequate emission should be used. But when slightly larger X-ray generating volume can be tolerated, the highest beam accelerating voltage available will give the best X-ray results.

This is because the generated X-ray flux increases approximately proportional to the amount by which the accelerating voltage exceeds the absorption edge of the element, since the incident electron has more energy and can ionize more atoms, producing more X-rays. Also, the peak-to-background ratio improves as the incident energy is increased, with the result that the figure of merit (I_o^2/I_B) improves and with it the minimum detectable limit. Table 3 illustrates this effect of accelerating voltage.

(There are many slightly different formulas used to calculate the minimum detectable limit. The one used for this table assumes that a peak is detectable above background if it rises above it by twice its standard deviation. Statistically this reduces to MDL $= 2\sqrt{2I_B}/I_o\sqrt{t}$ where I_o and I_B are measured with a pure element standard. Absorption and other effects are not considered here but are treated in a more detailed derivation by Ziebold [11].

For quantitative analysis it is often necessary, as described before, to calculate the magnitude of one peak for an element from another peak, and this requires their ratios to be constant. Measurements show, however, that the $K\alpha/K\beta$ and $L\alpha/L\beta/L\gamma$ ratios vary with incident beam energy. Table 4 shows that these ratios level off to a constant value, different for different elements but not dependent on accelerating voltages for voltages above about 2.5 times the absorption edge. This is probably due

TABLE 3—*Effect of accelerating voltage on detectability.*[a]

Aluminum $K\alpha$ 1.49 keV

kV	5	10	15	20	25
cps/pA	0.7	2.3	4.0	5.8	7.6
P/B	24	85	140	175	160
MDL[a] (%)	0.690	0.202	0.120	0.089	0.081

Titanium $K\alpha$ 4.51 keV

kV	5	10	20	30	40
cps/pA	0.1	2.3	7.1	11.9	16.6
P/B	1.1	46	88	112	135
MDL[a] (%)	8.53	0.275	0.113	0.077	0.060

Copper $K\alpha$ 8.04 keV

kV	10	20	30	40
cps/pA	0.3	4.0	7.8	11.6
P/B	2.7	46	81	104
MDL[a] (%)	3.14	0.209	0.113	0.081

[a] For 100-s counting time with 100 pA current.

TABLE 4—*Effect of accelerating voltage on line ratios.*

Titanium (K_{abs} = 4.96 keV)

kV	7.5	10	15	25	30	40
$K\alpha/K\beta$	12.15	8.77	8.64	8.59	8.56	8.55

Copper (K_{abs} = 8.98 keV)

kV	15	20	25	30	40	50
$K\alpha/K\beta$	9.36	5.27	5.16	5.10	5.08	5.08

Molybdenum (K_{abs} = 20.0 keV)

kV	25	30	35	40	50
$K\alpha/K\beta$	13.4	7.63	6.02	5.71	5.64

Gold (L_{abs} = 14.35 keV)

kV	15	25	35	45
$L\alpha/L\gamma$	10.98	8.02	7.92	7.89
$L\alpha/L\beta$	2.30	1.67	1.58	1.53

to the slightly different absorption of the different energy lines as the depth of electron penetration and the length of the X-ray path through the specimen change with incident electron energy. For both consistent line ratios and adequate emission, accelerating voltages of two to three times the absorption edge should be used whenever possible.

Beam Current

The SEM usually is operated at beam currents of 10^{-11} or 10^{-12} A for best image resolution in the secondary electron image. ED systems designed for SEM applications take advantage of the efficiency available and give satisfactory count rates at the lowest operating current. The count rate and speed of analysis can be improved by increasing beam current, but only up to the point where dead time becomes significant (typically about 10^{-9} A). It is also important to check the system calibration to be sure that gain does not shift with count rate.

Energy Resolution

The most commonly quoted specification of ED system performance is the full-width-at-half-maximum (FWHM) resolution. This value can give some idea of the ability of the system to separate peaks from adjacent elements when both elements are present in similar amounts. However, another common situation is the presence of a trace amount of one element in the presence of a large amount of its neighbor. This requires that the peak shape be uniform right down to the baseline. A perfectly Gaussian peak would have a full-width-at-tenth-maximum (FWTM) equal to 1.83 times FWHM, but improperly adjusted electronics (refer to paper in this volume by Walter) can cause spreading at the "skirts." Figure 14 shows the necessity of having "clean skirts" to detect, in this case, 0.65 percent manganese in steel.

Resolution Versus Detector Size

As described before, the larger the solid angle covered by the detector, the higher the count rate and the better the speed of data collection and detectability. Since the mechanical design, distance from specimen to detector, and takeoff angle usually depend on the size of the housing (which is much larger than the actual detector), it should be possible to use quite large area Si(Li) detectors. However, as discussed in the paper in this volume by Sandborg, increasing the area of the detector increases the input capacitance on the preamplifier and results in poorer resolution.

Table 5 compares the resolution, peak-to-background ratio and size of detectors. This indicates that for applications were detectability limits are

TABLE 5—*Performance of detectors versus size and resolution.*

Detector area.............	12.5 mm²	30 mm²	80 mm²
System resolution[a]........	225 eV	250 eV	300 eV
P/B ratio[b]..............	81	43	19
Area P/B...............	1012	1290	1520

[a] Approximate system resolution expected for comparable quality Si (Li) diode.
[b] Measured on copper $K\alpha$-peak with systems of indicated resolution.

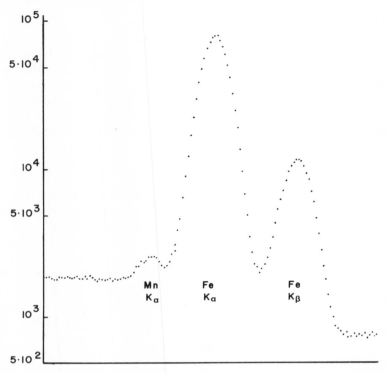

FIG. 14—*ED X-ray spectrum of steel containing 0.65 percent magnanese.*

important the figure of merit (I_o^2/I_B), which is proportional to (area X P/B ratio) can be improved somewhat by using the largest detector available. This would be particularly desirable, for instance, for the detection of trace elements in biological tissue. Mineral or metal studies, on the other hand, generate more X-rays and require the analysis of adjacent elements for which better resolution would be desired at some cost in count rate.

Selection

The selection of an ED system for an SEM should pay particular attention to several parameters, not all independent: resolution (FWHM and FWTM); solid angle subtended (a function of detector size *and* mechanical design); takeoff angle; gain stability (over long periods of time and over a wide count rate range; and low energy X-ray detection (determined by entrance window thickness). Other less technical, but still important, specifications include liquid nitrogen holding time, availability of multiple memory segments for spectrum comparison, and general ease of operation.

Conclusion

For application to the SEM the energy dispersion (ED) X-ray spectrometer has many advantages over wavelength dispersion (WD) units. The most important is high efficiency because of the large solid angle of coverage, inherent detector efficiency, and simultaneous processing of different energy X-rays. Also, the ED analysis is not affected by the rough specimens normally examined in the SEM and works at both lower and higher magnifications than the WD spectrometer. The ED spectrometer is used for the same types of analysis available with WD spectrometer or electron probe microanalyzers, including point analysis, line or modal analysis, X-ray area images, trace element detection, and quantitative analysis. The chief limitation is its inability to detect light elements. The ease of operation, interpretation, and maintenance are also important to many SEM operators.

References

[1] Kimoto, S. and Russ, J. C., "The Characteristics and Applications of the Scanning Microscope," *Materials Research and Standards,* Vol. 9, No. 1, pp. 8-16.
[2] Elad, E., Sandborg, A. O., Russ, J. C., and Van Gorp, T., "The Use of an Energy Dispersive X-ray Analyzer in Scanning Electron Microscopy," presented at *IEEE 1969 Nuclear Science Symposium,* San Francisco, Oct. 1969.
[3] Sutfin, L. V. and Ogilvie, R. E., "A Comparison of X-ray Analysis Techniques Available for Scanning Electron Microscopes," *Proceedings of the Third SEM Symposium,* Chicago, April 1970.
[4] Specimen courtesy of K. Ohmae, Massachusetts Institute of Technology.
[5] Russ, J. C. and Kabaya, A., "Use of a Nondispersive X-ray Spectrometer on the Scanning Electron Microscope," *Proceedings of the Second SEM Symposium,* Chicago, April 1969.
[6] Dorfler, G. and Russ, J. C., "A System for Stereometric Analysis with the Scanning Electron Microscope," *Proceedings of the Third SEM Symposium,* Chicago, April 1970.
[7] Tousimis, A. J., "Election Probe Microanalysis of Biological Structures," *X-ray and Electron Probe Analysis in Biomedical Research,* Plenum, New York, 1969.
[8] Russ, J. C. and McNatt, E. C., "Copper Localization in Cirrhotic Rat Liver by Scanning Electron Microscopy," *Proceedings of the Electron Microscope Society of America,* Sept. 1969.
[9] Russ, J. C., "Rapid Quantitative Bulk Analysis of Brasses Using the Energy Disperson X-ray Analyzer," presented at *Electron Probe Analysis Society of America,* New York City, July 1970.
[10] *X-ray Emission Line Wavelengths and Two-Theta Tables, ASTM DS 37,* American Society for Testing and Materials, 1965.
[11] Ziebold, T. O., *Analytical Chemistry,* Vol. 39, No. 8, 1967, p. 858.

S. L. Bender[1] and R. H. Duff[1]

Energy Dispersion X-ray Analysis with the Transmission Electron Microscope

REFERENCE: Bender, S. L. and Duff, R. H., "**Energy Dispersion X-ray Analysis with the Transmission Electron Microscope,**" *Energy Dispersion X-ray Analysis: X-ray and Electron Probe Analysis, ASTM STP 485,* American Society for Testing and Materials, 1971, pp.180–196.

ABSTRACT: The addition of X-ray emission analysis capabilities to a conventional transmission electron microscope offers a new analytical dimension to the microscopist.

The availability of Si(Li) solid-state X-ray detectors with energy resolutions superior to any previous type of X-ray detector provides the microscopist with an efficient system for elemental analysis that does not interfere with normal microscope operation.

A commercially available Si(Li) detector has been installed on a transmission electron microscope capable of forming a 1.6-μm-diameter spot at the specimen plane with the following characteristics of operation (*a*) X-radiation from the specimen, the specimen holder, the mounting grid, and the objective aperture (if in place) are the only signals detected; (*b*) no degradation of detector resolution by the magnetic fields of the microscope lenses or the proximity of the incident electron beam to the detector (a few millimeters) is observed; and (*c*) X-ray signals from particles 1000 Å in diameter are detectable with peak-to-background ratios of about two.

KEY WORDS: transmission, electron microscopy, X-ray spectroscopy, semiconductor devices, extraction, replicas, thin films, particle beams, electron diffraction, solid state devices, X-ray analysis, scattering, dispersing, resolution, spectrochemical analysis

The advantages of collecting and analyzing X-ray emission intensities from a specimen while observing its electron scattering properties with a transmission electron microscope (TEM) have been discussed by several authors [1-6].[2] Both Bragg dispersion [1,2,4,6] and energy dispersion [5] X-ray systems have been proposed and successfully utilized in conjunction with transmission microscopes. In addition to the energy dispersion X-ray system of Ref 5, Siemens America, Inc., has had a flow proportional detector with cathode ray tube or pulse height selector output available for

---

[1] Research chemist and analytical chemist, respectively, Ledgemont Laboratory, Kennecott Copper Corp., Lexington, Mass. 02173.

[2] The italic numbers in brackets refer to the list of references appended to this paper.

use with their transmission microscopes since approximately 1963. The detector is of the side window type which views the emitted X-ray beam above the specimen through a hole in the objective lens which was designed to accommodate the stereo drive.

The availability of Si(Li) solid-state X-ray detectors with resolutions better than 300 eV full width at half maximum (FWHM) at 6.4 keV provides the microscopist with an efficient system for elemental analysis permitting the resolution of the K-spectra of adjacent atomic number elements down to phosphorus and with windows capable of transmitting approximately 10 percent of incident 1 keV X-rays. Thus, high resolution morphological data, crystal structure data, and an elemental analysis can be obtained from a unique submicron specimen.

The problem at hand is to mount such a device on a transmission microscope such that a satisfactory compromise is reached between normal microscope operation and adequate X-ray emission spectrographic performance.

Unlike the electron probe or scanning electron microscope which, in general, have large specimen chambers that are relatively empty, the specimen chamber in a transmission microscope is confined to a mechanically complicated region of the objective lens. The magnetic lenses used in transmission microscopes for high magnification consist of two concentric pole pieces with a small gap between them. Short focal length objective lenses necessary for high magnifications require a high magnetic field strength, and thus the diameter of the pole piece bore must be kept small relative to the length of the gap. To maintain focus, the objective must be close to the first focal plane. The stringent geometric requirements necessary for a high resolution transmission electron microscope limit the accessibility to the emitted X-ray beam and present numerous sources of unwanted primary and secondary X-rays.

Solid-State Detector, Transmission Electron Microscope Interface

Figure 1 is a schematic diagram of our current instrument illustrating the detector to specimen configuration. The specimen holder is in the form of a cone with 11-deg half-cone angle which defines the X-ray beam emitted by the specimen under electron excitation. The detector cryostat assembly is inserted through the microscope cold trap port making a vacuum seal with a nylon flange (see Fig. 2). Nylon is used to electrically isolate the cryostat from the microscope. The flange is 1½ in. thick in order to allow movement of the cryostat in a lateral direction for proper alignment relative to the electron beam and for radiation protection. With the microscope operating at 50 kV and 20-μA beam current, no radiation is detected at the flange surface, and at 100 kV and 20 μA the radiation level is 3 mR/h. The maximum acceptable dose to the eyes is 30 mR/week.

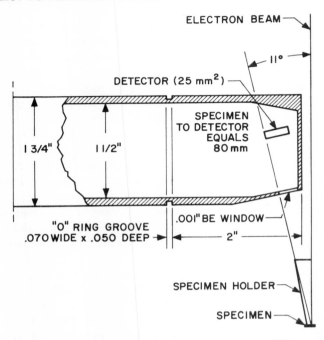

FIG. 1—*Schematic representation of Si(Li) detector, TEM interface.*

The diameter of the detector window is critical in keeping the angle of view of the Si(Li) crystal confined to radiation from the specimen surface. The optimum diameter for the window is that of the circular cross section at the window plane of the cone described by the point of specimen electron beam contact and the Si(Li) crystal. The 1 mil beryllium window is 6 mm in diameter (slightly larger than the optimum diameter). The 25-mm² Si(Li) crystal is 80 mm from the specimen and thus subtends 0.004 steradians.

In practice, with no additional collimation, the Si(Li) crystal intercepts X-radiation from the specimen, specimen grid, specimen holder, and molybdenum objective aperture, if in place. Additional specimen holders have been machined from aluminum and high-purity graphite from which no detectable X-radiation is emitted. Aluminum, nickel, copper, titanium, gold, and silver specimen support grids are readily available. Therefore, with judicious choice of specimen support grid, a graphite specimen holder and the objective aperture removed for the X-ray analysis, only radiation from the specimen and specimen grid are detected. Micrographs may be taken with the aperture in place either before or after the X-ray analysis.

Since the Hitachi specimen changing device must be mechanically positioned past the column center during change, it cannot be used with our present design. Specimens are changed through the lead glass window facing the operator and above the specimen chamber (see Fig. 2), and this must be done with the column at atmospheric pressure. Two solutions to this problem are possible: (1) attach mechanical stops to the cryostat can so that it can be moved back for specimen exchange and then precisely repositioned by the stop outside the column, and (2) design a specimen changing device that does not move past the column center line. The second solution is appealing and such a device has been designed and will be installed shortly.

Measurements of detector resolution at 6.4 keV with a Co^{57} source and iron $K\alpha$-radiation excited in the transmission microscope gave identical results. The FWHM at 6.4 keV for the Nuclear Diodes, Inc., detector is 270 eV. Therefore, the magnetic fields of the microscope lenses and the electron beam have no effect on the detector resolution of a 270-eV system.

FIG. 2—*Nuclear Diodes, Inc., Si(Li) detector on Hitachi HU-11A TEM.*

Performance Characteristics

The Hitachi HU-11A transmission electron microscope is routinely operated at 100 kV with beam currents in the range of 1 to 20 μA (beam currents in the μA range are common to most TEM). All the electron micrographs and electron diffraction patterns described in this paper were taken at 100 kV. The X-ray data were collected at 50 kV, since, in general, peak-to-background ratios were observed to be higher than at 100 kV for X-ray emission lines in the 1 to 10 keV range. This is not surprising since the electron cross section for continuum generation increases with increasing excitation potential while the electron cross section for characteristic generation begins to decrease at high overvoltage values depending upon the characteristic energy and atomic number of the scatterer. Since 50 kV is the lowest obtainable with the Hitachi HU-11A, we do not have data at lower accelerating potentials, but it is felt that lower poentials, in the range of 15 to 30 kV, would be an advantage for the light elements in particular. The beam currents used to produce electron micrographs and electron diffraction patterns produce X-ray intensities that swamp the capabilities of the solid state detecting system. This is true down to the smallest specimens studied, that is, 1000-Å-diameter particles. In order to operate at the low beam currents required for X-ray analysis and keep the filament in a saturated condition a continuously variable gun bias control was installed. The X-ray spectra were collected with beam currents lower than 1 μA except for the case of extremely small particles where 1 μA was used in order to keep data collecting times in the region of 100 s.

Dispersed particles of high-purity silicon carbide (SiC), chrominum boride (Cr_2B), nickel boride (Ni_2B), and zirconium boride (ZrB_2) were studied to obtain mass sensitivity information for a large emission energy range (1.74 to 15.7 keV) with simple specimens. The carbide and borides were selected for their spectral simplicity and lack of large secondary X-ray interactions such as absorption and fluorescence, that is, for silicon $K\alpha$ in carbon μ/ρ equals 360 which is the worst case. Figure 3 consists of X-ray spectra from SiC particles dispersed on a carbon film placed on a nickel grid. The electron beam is defocused to 10 μm in diameter. The first spectrum is from five SiC particles with a combined volume of approximately 10 μ^3. Except for the nickel K-spectrum from the grid and the silicon K-spectrum (integrated intensity equals 230 counts per second) the background is smooth to 10 keV (no stray emission lines appear to 50 keV). The second spectrum is from four SiC particles with a combined volume of approximately 1μ^3. The third spectrum is taken of SiC particles at the center of a grid opening indicating that the grid material is excited by high energy scattered electrons. The X-ray spectra of Cr_2B and Ni_2B particles on a titanium grid are displayed in Fig. 4. Spectra 4 and 5 are from Cr_2B particles with combined volumes of 2 and 14 μ^3, respectively

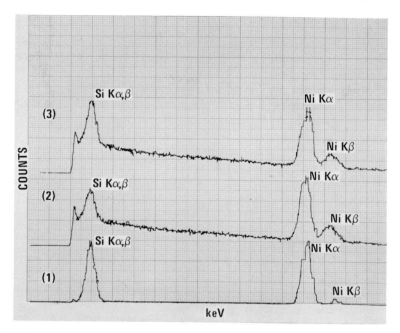

FIG. 3—*X-ray spectra of SiC particles on nickel grid, graphite holder.*

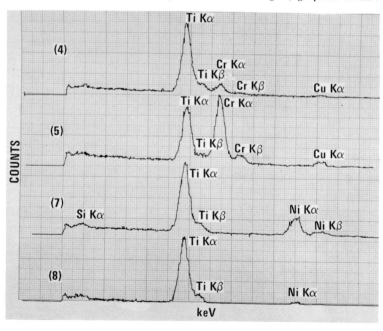

FIG. 4—*X-ray spectra of Cr₂B and Ni₂B particles on titanium grid, graphite holder.*

(the Cr_2B apparently contains copper as an impurity and possibly silicon). Spectrum 7 is from a single Ni_2B particle 2.5 μm in diameter (8 μ^3) and Spectrum 8 from a Ni_2B particle 1 μm in diameter. The peak-to-background ratio of the nickel $K\alpha$-line in Spectrum 8 is approximately 20 (the electron beam diameter was 1.6 μm). Figure 5 is the spectrum from a 7-μm-diameter particle of ZrB_2 on a titanium grid. With the electron beam focused (1.6 μm diameter) a peak-to-background ratio for the zirconium $L\alpha,\beta$ of 1.7 is observed for a 1000-Å-diameter particle of ZrB_2.

Applications

The following applications were selected as being representative of the types of specimens encountered by electron microscopists as well as illustrating the advantages of X-ray emission analysis. They consist of the examination of a massive specimen (no electron transmission except at the edges), a thin film, dispersed particles, and an extraction replica.

Massive Specimen

The specimen was removed from a graphite mold and placed unsupported in a graphite specimen holder. Although it was generally too thick to transmit electrons, diffraction was possible at some of the edges. Figures 6a and b show the electron diffraction pattern and the X-ray emission spectrum from the same area.

The diffraction pattern is identified as that belonging to the wurtzite structure zinc sulfide (ZnS) with $a_o = 3.2$ Å and $c_o = 5.2$ Å. Of the known compounds with the wurtzite structure, only zinc oxide (ZnO), $a_o = 3.25$ Å, $C_o = 5.21$ Å, and cobalt oxide (CoO), $a_o = 3.21$, $c_o = 5.24$ are consistent with the diffraction data. Since electron diffraction photographs are accurate in d-spacing to ± 2 percent, diffraction data alone cannot differentiate between ZnO and CoO. The X-ray emission spectrum clearly confirms the presence of ZnO. The small peak on the low energy side of the zinc $K\alpha$-peak indicates the presence of a small concentration of

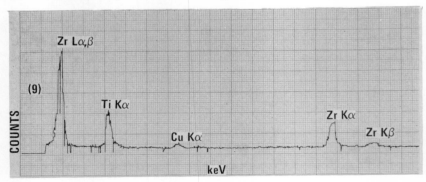

FIG. 5—*X-ray spectrum of ZrB_2 particles on titanium grid, graphite holder.*

FIG. 6a—*Electron diffraction pattern of massive specimen from graphite mold identified as ZHO.*

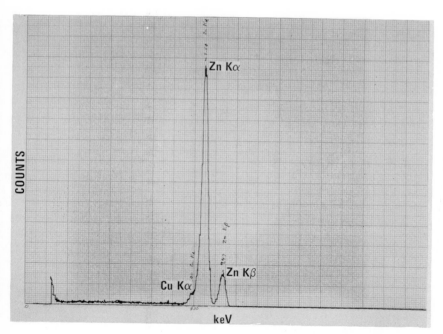

FIG. 6b—*X-ray spectrum of massive specimen, unsupported, graphite holder.*

copper. Additional electron diffraction patterns taken close to this site showed the presence of a small amount of copper oxide (CuO). The ratio of the zinc $K\alpha$ to copper $K\alpha$ peak heights (~20 to 1) agrees with the weight ratio of zinc to copper obtained by chemical analysis. The positive identification of zinc oxide inclusions by a combination of electron diffraction and X-ray emission data is a classic example of the solution of the ambiguity of diffraction data caused by isomorphs. Electron diffraction techniques are subject to an additional ambiguity due to the lack of reliability of electron diffraction intensities. Two compounds with different structures but the same Bravais lattice with close or identical lattice constants may be differentiated with X-ray diffraction intensities but not, in general, with electron diffraction intensities. Such problems can usually be resolved wth X-ray emission data.

Thin Film

A lead sulfide film (100 Å thick) was evaporated onto a sodium chloride substrate, floated off on water, picked up on a nickel grid, dried, and mounted in a copper specimen holder. Figures 7a, b, and c give the diffraction pattern, micrograph, and X-ray spectrum of this film. The copper signal is from the copper specimen holder and the nickel signal from the nickel grid.

FIG. 7a—*Selected area electron diffraction pattern of PbS thin film.*

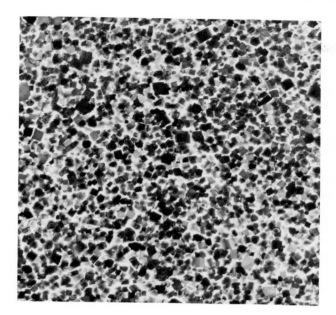

FIG. 7b—*Electron micrograph of PbS thin film, ×20,000.*

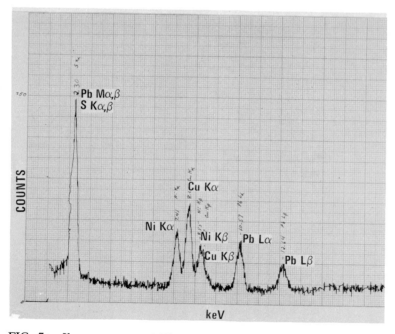

FIG. 7c—*X-ray spectrum of PbS thin film, nickel grid, copper holder.*

The spectrum of Fig. 8 is from a lead foil. The lead $M\alpha,\beta$-line is approximately 1.7 times the intensity of the lead $L\alpha$. One must have knowledge of this ratio as well as of changes in the lead M to L intensity ratios as a function of specimen thickness due to self absorption effects in order to interpret the apparent four to one intensity ratio between the lead $M\alpha,\beta$ and lead $L\alpha$ as due to the presence of sulfur $K\alpha,\beta$. Here is a typical example of interference where the energy difference between lead $M\alpha,\beta$ and sulfur $K\alpha,\beta$ is approximately 40 eV, which is well below the resolution of our system as well as current state of the art systems.

Dispersed Particles

Particles from a blocked filter were dispersed ultrasonically in water, and a drop of this suspension was allowed to dry on a carbon film suspended on a nickel grid. Figures 9a and b show the electron diffraction pattern and the X-ray emission spectrum from an agglomeration of particles approximation 1 μm in diameter. Silicon and sulfur are present; however, iron and tin are the predominant elements detected. The electron diffraction pattern is identified as $Sn,Fe(O,OH)_2$, and thus the data are consistent. Silica is present as a-quartz, as identified in other electron diffraction patterns. The sulfur (or possibly the lead M-line, since lead was found in trace amounts) spectrum indicates a low concentration and is possibly in solution with the primary tin-iron phase. Since the X-ray emission spectrum arises from a larger volume than does the electron diffraction pattern,

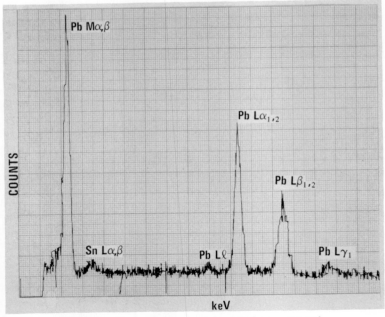

FIG. 8—*X-ray spectrum of lead foil unsupported.*

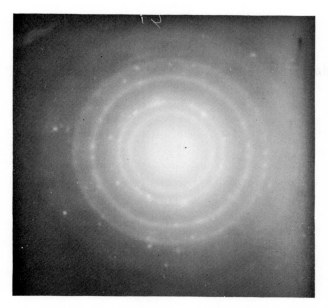

FIG. 9a—*Selected area electron diffraction pattern of filter residue identified as Sn,Fe(O,OH)₂.*

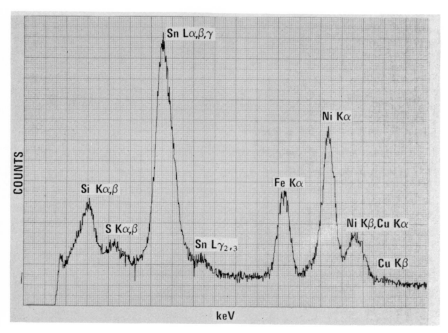

FIG. 9b—*X-ray spectrum of filter residue, nickel grid, graphite holder.*

all elements present in the spectrum need not necessarily be accounted for in the phase determinations.

Extraction Replica

An extraction replica was prepared by mechanically stripping a collodion replica from a copper bar in the vicinity of visually observable scale. The replica was carbon coated, the collodion dissolved away, and replica placed in the microscope on a titanium grid. Figure 10 shows the electron diffraction patterns, electron micrographs, and X-ray emission spectra from four extracted particles. The diffraction pattern corresponding to the first spectrum and the fine particles in the micrograph were identified as principally Cu_2O and $CuFeO_2$. The presence of these two phases in the scale was verified by X-ray diffraction. Diffraction patterns of the large piece seen in the micrograph reveal the presence of silica.

Diffraction patterns corresponding to the second and third spectra are identified as Cu_2O. Morphologically, the agglomerates are similar, and similar to the fines of the first micrograph.

The electron diffraction pattern corresponding to the last spectrum is a single crystal pattern which is not inconsistent with that of $CaSiO_3$. The fines seen in the micrograph are probably similar to those above, that is, Cu_2O which accounts for the copper signal.

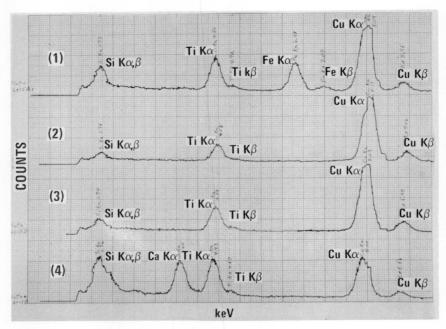

FIG. 10a—X-ray spectra of extracted particles.

FIG. 10*b*—*Selected area diffraction pattern of fine particles identified as* Cu_2O *and* $CuFeO_2$.

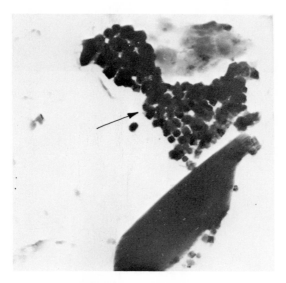

FIG. 10*c*—*Electron micrograph of fine particles.*

Discussion

The problems, design, operating characteristics, and applications discussed are closely related to those that will be encountered in an attempt to interface a solid-state X-ray detector with an old or newly purchased transmission electron microscope. One will generally be faced with a narrow X-ray beam that is difficult to get at and many sources of unwanted primary and secondary X-radiation. Once the problem of getting the detector cryostat into the column such that the Si(Li) crystal can intercept the unobstructed X-ray beam emitted by the specimen is solved (this appears to be possible with a majority of commercial TEM with or without a compromise such as our elimination of the specimen chamber cold trap),

FIG. 10d—*Selected area diffraction pattern of agglomerate identified as Cu_2O.*

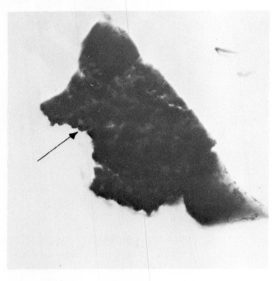

FIG. 10e—*Electron micrograph of agglomerates.*

the quality of the spectrum becomes the prime consideration. Only X-radiation from the specimen and possibly one or two additional sources with controllable emission characteristics can be tolerated. Two prime considerations in designing the cryostat are window size and window to Si(Li) crystal distance. The Si(Li) crystal should be as far as possible from the window so that it will only detect X-rays that enter the window at close to

FIG. 10f—*Selected area diffraction pattern of single crystal CaSiO₃.*

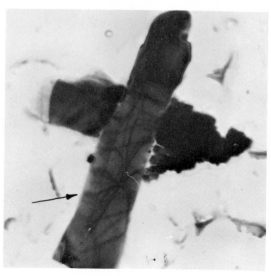

FIG. 10g—*Electron micrograph of CaSiO₃ single crystal.*

90 deg incidence. This also can be accomplished with a collimator, but it is more difficult to design and position one in the confined quarters of the specimen chamber. The window should not be larger than the projection of the Si(Li) crystal described by the point source of X-rays emitted by the specimen.

Now that Si(Li) detectors are available with resolutions at 6.4 keV of better than 200-eV FWHM, and since the TEM can take advantage of the best resolutions available due to the high X-ray intensities generated, energy dispersion X-ray analysis with the transmission electron microscope should become commonplace. It is hoped that future microscopes will be designed with energy dispersion X-ray capabilities in mind.

References

[1] Duncumb, P., "Electron Microscopy," *Proceedings of the Fifth International Congress for Electron Microscopy,* Vol. I, KK-4, Academic Press, New York, 1962.

[2] Akahari, H., Katagiri, S., Ozasa, S., and Fujiyosu, T., "Electron Microscopy," *Proceedings of the Sixth International Congress for Electron Microscopy,* Vol. 1, Maruzen Co., Ltd., Tokyo, Japan, 1966, pp. 187-88.

[3] Duncumb, P., *The Electron Microprobe,* Wiley, New York, 1966.

[4] Ogilvie, R. E., Moll, S. H., and Shippert, M. A., *Proceedings of the Twenty-Sixth Annual Meeting,* Electron Microscopy Society of America, Claitor's Publishing Division, Baton Rouge, La., 1968, pp. 310-311.

[5] Duff, R. H. and Bender, S. L. *Proceedings of the Twenty-Seventh Annual Meeting,* Electron Microscopy Society of America, Claitor's Publishing Division, Baton Rouge, La., 1969, pp. 130-1.

[6] Cooke, C. J. and Openshaw, I. K., *Proceedings of the Fourth National Conference on Electron Microprobe Analysis,* Pasadena, Calif., 1969.

L. V. Sutfin[1] *and R. E. Ogilvie*[1]

Role of the Gas Flow Proportional Counter in Energy Dispersive Analysis

REFERENCE: Sutfin, L. V. and Ogilvie, R. E., "**Role of the Gas Flow Proportional Counter in Energy Dispersive Analysis**," *Energy Dispersion X-ray Analysis: X-ray and Electron Probe Analysis, ASTM STP 485,* American Society for Testing and Materials, 1971, pp. 197–216.

ABSTRACT: A comparison of the gas flow proportional counter and the Si(Li) solid-state detector is made. Currently the flow proportional counter provides a means of extending the range of energy dispersive analysis to the region below 1 keV. It also can be used where the rigid environmental requirements of solid-state detectors cannot be met. The inherent statistical fluctuations of the ionization processes involved determine the ultimate resolution of gas counters, and this is inferior to that of the Si(Li) detector. Since energy analysis in the low energy region is more difficult than in the region above 1 keV, the solid-state detector appears to be worth developing for operation over the entire energy range.

KEY WORDS: gas flow proportional counters, chemical analysis, X-ray analysis, X-ray fluorescence, electron probes, solid state counters, evaluation, tests, dispersing, resolution, spectrometers

Like the semiconductor detector, the gas flow proportional counter detects radiant energy by converting it into an electrical pulse through the process of ionization. This device is one of a family of gaseous ionization detectors which includes the ionization chamber and Geiger counter. They were first developed to detect nuclear radiations and were later adapted to use in detecting hard, soft, and then ultra-soft X-rays.

Flow proportional counters have been used extensively in microprobe and X-ray fluorescence analysis using wavelength dispersive techniques. The output of the gas flow proportional counter is proportional to the energy of the photon absorbed; thus, electronic discrimination can be used to enhance the performance of wavelength dispersive spectrometers by eliminating unnecessary background and by separating higher order interferences. However, the energy discrimination of gas flow proportional

---

[1] Post-doctoral fellow and professor of metallurgy, respectively, Massachusetts Institute of Technology, Cambridge, Mass. 02139.

counters is often not good enough to completely resolve higher order interferences when one element is present in trace amounts [1].[2]

Before the utilization of pseudocrystals of the Langmuir-Blodgett type,
which extended the range of wavelength dispersive spectrometers down
through the lightest elements, energy dispersive analysis appeared attractive for light element analysis. Energy dispersive analysis with gas flow
proportional counters also has received some attention because relatively
large solid angles and high conversion efficiencies are possible. However,
the necessity for spectrum unfolding makes even qualitative analysis quite
complicated [2,3].

Precision always is affected adversely when spectrum unfolding techniques are used since variances are increased by mathematical manipula-
tion. As will be discussed later, the amplitude of the output pulse from a
gas flow proportional counter is usually a function of the rate of absorption of the incident photons as well as their energy. This characteristic of
the counter complicates spectrum unfolding and can reduce both accuracy
and precision.

The rapid development of Si(Li) detectors over the past few years has
generated a high level of interest in energy dispersion techniques which is
justifiable in view of the superior performance of this detector over that
of the proportional counter. Although the cost of acquiring and maintaining such a system is much lower than that of Si(Li) detectors, they are
not competitive for the reasons listed above. This is made manifest by the
fact that energy dispersion analysis with the proportional counter never
gained wide acceptance, whereas Si(Li) detectors in spite of their high
cost are being used widely and with increasing frequency.

If the range of Si(Li) detectors covered the light elements while still
exhibiting the inherently superior energy resolutions, the proportional
counter would not enter into energy dispersion analysis today. At present,
the Si(Li) detectors have a practical lower limit of 1 keV, and the proportional counter represents a way of extending the range of energy dispersion
analysis to low energy photons.

However, the practicality of this technique with either detector below 1
keV is somewhat dubious because of the plethora of X-ray energies which
exist in this range. For example, in the range of 200 eV on either side of
the iron $K\alpha$ (6.403 keV), Bearden's tables [4] list 29 possible X-ray energies; however, only two common elements produce photons with energies
in this range, iron and manganese. In contrast, in a range of energies 100
eV above and below the carbon $K\alpha$ line (277 eV), 87 possible lines of L,
M, N, O series are listed, and practically all of these are from common
elements. Even if adjacent elements in this range could be resolved, the

---

[2] The italic numbers in brackets refer to the list of references appended to this
paper.

problem of interpreting spectra is going to be more difficult than in higher energy ranges because of interferences from heavier elements.

Energy dispersive analysis of low energy photons may prove useful in analysis where the number of possible elemental combinations is restricted sufficiently, such as in organic materials where the problem may be limited to resolution of carbon, nitrogen, and oxygen K photons and possibly photons from one or two heavier elements. In biological applications the problem is already quite complex since within ± 100 eV of C_k energy one finds chlorine, potassium, and calcium L-lines, and osmium and lead N-lines. These five elements, all of which are important to the biological investigator, are found within an energy range which is smaller than that in which one might expect to find the C_k pulse height distribution.

Since carbon, nitrogen, and oxygen are not trace elements in organic or biological chemistry, energy dispersive analysis may prove useful. However, in fields such as metallurgy, where light elements are usually trace elements, it appears that this technique of analysis of light elements may not become practical.

Another difficulty which will be encountered in microprobe analysis is that the number of photons in the continuous spectrum generated by the electron beam is very high at low energies. Carbon is particularly difficult to measure at accelerating voltages greater than a few kiloelectron volts [2,5]. Consequently energy dispersive analysis for carbon must be done at energies too low to excite heavier elements, such as iron, and simultaneous analysis for these two elements may not be possible.

In discussing the characteristics of proportional counters we shall emphasize the detection of energies below 1 keV.

Detector Configuration

In general a gas flow proportional counter consists of an anode and cathode separated by a volume of detector gas. Counters which have entrance windows thin enough to transmit photons with energies of a few hundred electron volts will not maintain the pressure and purity of the detector gas; therefore, the detector gas is allowed to flow through the counter at a rate sufficient to maintain the desired pressure wihin the detector and flush out impurities which enter the active volume.

Two configurations have been used: the point anode detector and the coaxial detector. The point anode type has been shown to have poorer resolution than the coaxial type [6], and its configuration does not permit it to be used in a tandem arrangement with other detectors. For these reasons it will not be considered further.

The coaxial detector shown in Fig. 1 consists of a cylindrical tube, which is the cathode, and a fine wire strung along the axis of the tube, which is the anode. An entry port in the side of the tube permits the X-radiation to enter the active volume. Another port located diametrically opposite the entry port permits the radiation not absorbed in the active volume of the detector to pass through without producing fluorescent radiation of the tube material. The ports must be covered by window materials which will transmit efficiently low energy photons.

Windows thin enough to transmit low energy photons also will transmit backscattered electrons; therefore, when used in microprobe applications, electrostatic or magnetic electron traps must be used to prevent electrons from entering the detector and increasing the background level. The design of such a trap must be such that the electron beam is not defocussed or made astigmatic by its presence.

Ports also must be provided for gas flow, and, if the detector is to be physically located in a vacuum chamber, the detector construction should provide leak tight seals to minimize the effect on the vacuum system. There has been a wide variety of designs described by different investigators and we wish to refer the reader to the specific reports for more details [6-9]. Certain aspects of the geometrical configurations are of essential importance, and they will be discussed further in relation to performance characteristics.

One of the important advantages of energy dispersive analysis is that the entire spectrum can be obtained by one detector, eliminating the difference in absorption path when more than one spectrometer is used. Although not absolutely necessary, it would be desirable to maintain this advantage. This can be done by mounting the proportional counter in tandem with the solid-state detector as shown in Fig. 2. Certain problems will be created, however. More space is required and the solid-state detector will be farther from the source, the intensity of radiation above 1 keV will be decreased since it must pass through the proportional counter, and the materials in the proportional counter may be excited and contribute to the spectrum obtained from the solid state detector.

Basic Phenomena

The proportional counter converts electromagnetic radiation to an electronic pulse through the ionization of the detector gas which fills the active volume of the detector. The counter also can amplify the pulse by a factor

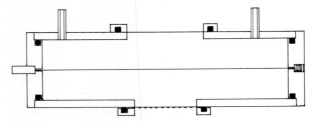

FIG. 1—*Coaxial gas flow proportional counter.*

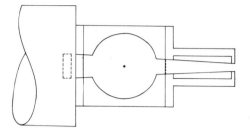

FIG. 2—*Tandem arrangement of gas flow proportional counter and solid-state detector with electron trap.*

as large as 10^4. The process of producing the electrical pulse can be considered as having three parts: conversion, energy absorption, and avalanche production. Conversion and energy absorption occur in the semiconductor detectors and ionization chambers, while all three occur in the proportional and Geiger counter. The avalanche produces the amplification of the pulse which occurs in the latter two devices. At the energies being detected in microprobe analysis and X-ray fluorescence the initial interaction between the incoming photon and the gas will be the production of a photo electron; the probability of Compton scattering and pair production being very low [10]. Ideally the initial ionization will be of the outer shell; however, inner shell ionizations do occur when the energy of the photon is great enough. Inner shell ionizations can be detrimental to the performance of the detector for reasons which will be discussed later.

The photo electrons freed will have an energy, E, equal to that of the initial photon, E_p, less the energy required for the ionization, E_i:

$$E = E_p - E_i$$

Since E_i for the gases used is on the order of 10 eV and the photon energies will be above 100 eV, E is large enough to cause further ionizations of the gas. Further ionizations will occur until no electron has sufficient energy to ionize the gas atoms. At this point the interaction of the photon with the detector gas has created a number of free electrons, $\overline{N}$, and positive ions which on the average is proportional to the energy of the absorbed photon:

$$\overline{N} = Ep/\epsilon$$

There are, however, competing processes which also absorb energy but do not produce a free electron. Atoms also will be excited to emit light, and heat also is created. Therefore, ϵ, the energy absorbed per ionization is significantly greater than the ionization potential. For example, for argon, a common detector gas, the average energy absorbed per electron collected is approximately 27 eV, whereas the ionization potential is 15.7 eV [11].

The electrons thus freed drift toward the central wire, and the ions drift toward the wall. The field present in the active volume of a coaxial detector is described by the following equation [11]:

$$E(r) = \frac{V}{r \ln r_a/r_c}$$

If the anode is quite thin compared to the inside diameter of the detector; that is $r_a \ll r_c$, the field near the anode becomes relatively large. If the detector voltage is sufficient, an electron can gain enough energy between collisions; that is, in one mean free path length, as it approaches the anode to create further ionizations of the detector gas.

For a fixed geometry, the threshold of gas amplification or avalanche production is considered to be the voltage, V, where an electron gains enough energy in the last mean free path before impinging on the anode to cause an ionizing event. As the potential is increased from this threshold value the amplification factor increases, and the total number of electrons collected at the anode can be many times greater than that number created by the initial ionizations. Because of this avalanche of electrons, the signal is amplified in the counter itself. Since no current flows in the detector when no radiation is being absorbed, the amplification is noiseless and pulses from low energy photons can be distinguished from electronic noise. Eventually amplification ceases to be constant and becomes dependent on the energy of the absorbed photon. Finally, the amplitude of the output pulse tends to become constant; that is, reaches a maximum, and the detector becomes a Geiger counter with no ability to discriminate.

The collection of the electrons at the anode results in an accumulation of charge

$$\bar{q} = \bar{N} \times e \times \bar{A}$$

which can be converted to a voltage across a capacitor.

$$V_c = \bar{N}e\bar{A}/C.$$

where

$\bar{N}$ = average number of electrons created by the initial ionization of the gas,
e = electronic charge,
$\bar{A}$ = average amplification factor, and
C = capacitance.

The voltage rise depends upon the rate at which the electrons are collected. However, the collection of positive ions, which because of their greater mass and collision cross section is a much slower process, also contributes to V_c increasing the total charge collecting time. In counters which record single pulses the capacitor is discharged through a resistor, and the effect of positive ion collection is a tailing of the pulse.

Another phenomenon is related to positive ion collection. As the ion approaches the cathode the local field strength can increase enough to cause an electron to leave the surface of the metal and neutralize the ion. The energy dissipated in this inelastic collision can result in an excited atom which will emit an ultraviolet photon. The photon can have enough energy to eject one or more electrons from the wall. These electrons can create a charge pulse just as those created by the absorption of an X-ray photon [13].

If a polyatomic gas which absorbs the ultraviolet photon efficiently is added to the primary detector gas, the ejection of electrons from the walls can be prevented because polyatomic molecules tend to undergo chemical change rather than fluorescing when in an excited state. This phenomenon of quenching becomes important at high amplifications and high count rates. The production of photoelectrons has a small probability (approximately 10^{-4}) and only becomes effective when the number of ions is great. The prolonged pulse which would occur if the discharge were not quenched would degrade the count rate capabilities of the detector [13].

It was mentioned earlier that inner shell ionizations are detrimental to the performance of the detector. This is because the atom in the ionized state may emit an X-ray photon in returning to ground state. (The probability of doing so is the fluorescence yield.) Since the absorption by a material of its own characteristic X-ray is relatively low, the probability that the photon generated within the detector will escape can be significantly large. The result of this escape is a pulse which is of a smaller voltage than expected. The difference in voltage is that which is equivalent to the energy of the photon which escaped. A bimodal distribution of the output results. This is particularly troublesome in energy dispersive analysis since it complicates spectrum unfolding. Bimodal distributions of this type occur only when E_p is greater than energy of the absorption edge of the detector gas.

Operating Characteristics

The performance of proportional counters can be described by evaluating the following characteristics: amplification, efficiency, range, stability, and resolution. Of these the first three can be quite adequate for energy dispersive analysis, but the last two are not good enough to allow the analysis to be performed without the employment of complicated data processing techniques. Proportional counters can be used to detect and discriminate ionizing radiation with energies ranging from a hundred or so electron volts to well over a million electron volts; however, the following discussion will emphasize the detection of low energy X-rays. Unfortunately those who have played a major role in the development of proportional counters and proportional counter theory have been interested in analysis of higher energy radiation encountered in nuclear physics. There has not been a good basic publication devoted to the detection of low energy photons. At this time it appears that such studies would be fruitful not only in extending energy dispersive analysis down to the lightest elements but also in improving the performance of wavelength dispersive X-ray spectrometers.

Amplification

At present one of the limitations of the solid-state detector is that the electronic noise is comparable to the pulse produced by low energy radiation. The feature of proportional counters which makes them of interest at all is that the detector itself can amplify a pulse several orders of magnitude; therefore, low energy pulses can be distinguished readily from the electronic noise.

A proportional counter can be made to function as an ionization chamber; that is, without amplification, and as a Geiger counter; that is, with full discharge of the tube regardless of the energy of the absorbed radiation. Figure 3 shows qualitatively the change in pulse amplitude as the anode voltage is increased. In the ion chamber region the field strength is not sufficient to accelerate an electron in one mean free path length enough to cause further ionization of the detector gas. In this region the output pulse is proportional to the energy of the absorbed photon and is not a function of the detector voltage. At a detector voltage V_p the pulse size begins to increase. As the voltage is increased the number of ionizing events in the avalanche increases. On the average the size of the avalanche is proportional to the number of initial ionizations, and the output pulse remains proportional to the energy of the absorbed photon but is an exponential function of the detector voltage.

In the limited proportional region the size of the avalanche is proportionately less for higher energy photons than for low energy photons as the avalanche begins to approach the maximum which exits in the Geiger region. Operation in the limited proportional region tends to compress the spectrum. If one were to operate in this region unknowingly, results could be easily misinterpreted. Proper calibration of the system should automatically reveal any nonlinearities.

That there is a limit to size of the avalanche is demonstrated by the existence of the Geiger region, where the average pulse size is not dependent on the energy of the absorbed photon but is constant and the creation of one electron can cause complete discharge of the tube. Increasing the voltage beyond the Geiger region results in continuous discharge of the counter.

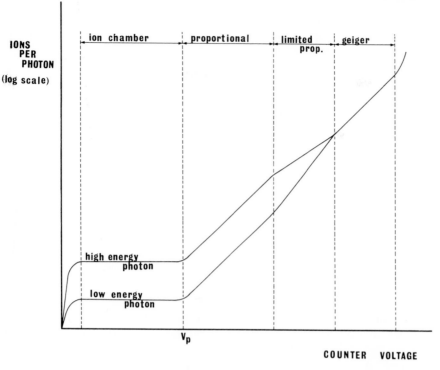

FIG. 3—*Ionization in counter as function of counter voltage.*

Several relationships have been developed for predicting the value of $\bar{A}$, the amplification factor of the counter. Regardless of whether these relationships are derived from theory or developed empirically, they are good only over a limited range of gas pressures and compositions. Because of this lack of generality it is best to refer the reader to the specific references for details [14-16].

The amplification factor for a coaxial type counter detecting X-rays is a function of the radii of the anode and cathode, the composition of the gas, the pressure of the gas, and the voltage applied. For any given counting system there will be a range of gas gains or counter voltages which will produce optimum resolution. The lowest value of gas gain in this range should be used to reduce space charge effects. The dependence of resolution on gas gain for a coaxial type counter is shown in Fig. 4. A similar dependence was reported by Williams and Sara [16].

Space charge effects are caused by the slow collection of positive ions. Any counting rate which is less than the rate of electron collection but greater than the rate of positive ion collection will result in a net accumulation of positive ions within the counter. The positive ions will tend to reduce the potential, reducing $\bar{A}$. Since X-ray counting rates are not regular

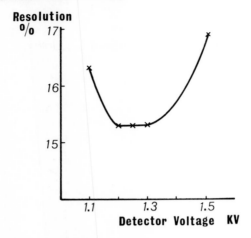

FIG. 4—*Optimum counter voltage range.*

with short intervals between quanta being more probable [*18*], then $\bar{A}$ will vary from time to time within a given counting period. Also, the average value of $\bar{A}$ will begin to decrease at a given count rate, which produces a significant space charge. It would be expected that $\bar{A}$ would not show significant changes at count rates below 10^4; however, the change in $\bar{A}$ with count rate can be continuous from very low count rates [*19*]. This will be discussed more fully in the section on stability.

Efficiency and Range

The range of operation of a proportional counter is limited by the absorption of the photon in the sensitive volume of the detector. At the low energy end of the range the limit is set by the transmission characteristics of the window. At the high energy end the limit is set by failure to be absorbed in the detector. Another factor which must be considered is the solid angle intercepted by the detector window. This is determined by the size of the window and the distance from the source. Since thin windows must be supported by grids, then the open area of the grid also will determine in part the solid angle intercepted.

Henke has published data that show the high energy cutoff; that is, 50 percent efficiency, for 2-cm-diameter counters operating at 1-atm pressure to be 2.48 keV for P-10 gas (90 percent argon–10 percent methane) and 1.24 keV for pure methane [*20*]. Varying the gas pressure and detector diameter will vary the high energy cutoff.

Thin windows are necessary if low energy photons are to enter the active volume. The transmission characteristics of the window are determined not only by its density and thickness but also by its chemical composition. Transmission curves for both stretched polypropylene and Formvar films have been published by Henke [*21*]. Pinhole-free polypropylene stretched by the vacuum method described by Caruso and Kin can be used [22]. It is an effective filter of nitrogen K-radiation and is only a poor transmitter of oxygen K-radiation, whereas it is relatively good for carbon, boron, sodium, magnesium, and aluminum K-radiation; that is, transmission > 70 percent.

Formvar films cast on water transmit nitrogen K- and oxygen K-radiation more efficiently because Formvar contains 32 percent oxygen, which transmits both oxygen K- and nitrogen K-radiation better than carbon, the major constituent of polypropylene. The nitrogen K-transmission of such films is given by Henke as approximately 45 percent [20].

It is conceivable that nitrogen K-transmission can be improved by using polymers which contain significant amounts of nitrogen, such as polymers of urea.

Also, by using less massive windows the absorption is reduced. Pinhole-free films of polyparaxylene can be produced in sizes large enough to serve as detector windows and thin enough to transmit greater than 80 percent of oxygen K-radiation and greater than 70 percent nitrogen K-radiation.[3] These films are not produced readily in the laboratory, as are Formvar and polypropylene films, but their absorption characteristics warrant the serious investigation of their use as detector windows. If windows with these transmission characteristics could be developed, then the efficiency over the range below 1 keV should be quite adequate.

Stability

It was mentioned earlier that space charge should reduce the size of the avalanche, resulting in a smaller pulse amplitude at higher counting rates than at lower ones. Positive ion collection time has been estimated to be of the order of 10^{-4} s, and one would not expect the pulse amplitude reduction to be significant at count rates less than several thousand counts per second (cps). Figure 5 shows the change in pulse amplitude with count rate for a proportional counter using P-10 (90 percent argon–10 percent methane) gas. The incident radiation was copper $K\alpha_1$. The change was continuous over the whole range measured.

Bender and Rapperport found greater changes, as large as 15 percent, in the peak position in several gas flow proportional counters over the range of 20 to 1000 counts per second. They also used monochromatic incident radiation but measured the effect at several different photon energies. The relative pulse amplitude was found to be more sensitive to the energy of the photon than to the value of gas gain. Figure 6 shows this effect. The (change in pulse height)/(change in energy absorbed) is greater for low energy photons than for high energy photons.

[3] Spivack, M. A., Union Carbide Corporation, Bound Brook, N. J., personal communication.

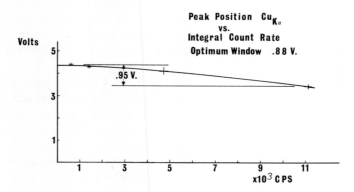

FIG. 5—*Change in pulse amplitude with countrate. The change is great enough to cause a significant loss of counts if an optimum window for detectability is used. Counter gas P-10.*

The dependence on photon energy is consistent with space charge being the mechanism creating the shift. Since the more energetic photons have a longer mean path in the detector, they will be absorbed in a larger volume. The charge density will not be as great for a given rate of energy absorption for the higher energy photons. Decreasing the intensity of the beam (energy/area/time) at a constant count rate also has been found to reduce the effect of count rate on pulse amplitude. This indicates that gas density should be lowered rather than the diameter being reduced in a detector designated to selectively absorb low energy photons to minimize the count rate effect.

Pulse amplitude shifts which occur in a gas flow proportional counter used in an energy dispersive system will depend on the energy distribution of the spectrum as well as the total count rate. The effect of energy absorption rate on the linearity of the spectral analysis should be investigated.

The effect on pulse amplitude at very low rates of energy absorption is not explained by assuming space charge effects due to ionization of the counter gas.

Speilberg has found the anode material and the condition of the surface to be important. He reported that heating the anode of the detector while under vacuum eliminated the dependency of pulse amplitude on count rate but with use the effect would return. He stated that a film is built up on the anode during use and that the contaminants were responsible for the pulse amplitude decrease [23-24].

There are four possible sources of contaminants: (1) the detector gas may contain impurities; (2) foreign gases may diffuse into the counter through the thin windows; (3) the polyatomic component undergoes degradation during use, and (4) materials used in the construction of the counter may give rise to impurities.

Culhane has reported that water vapor diffusing through the thin window of a gas flow proportional counter has been found to decrease performance [25]. Korff discusses the effect of use on the counter gas. He states that methane can polymerize during use and other polyatomic gases disintegrate. He states that the contaminants are created by radiation damage of the gas and their deposition on the wire degrades performance [13].

Figures 7 and 8 are presented as evidence that materials used in the construction of a detector can be a source of contaminants. Figure 7 is a scanning micrograph of a new anode wire and one which has been used until its performance degraded. The detector was a sealed counter filled with xenon and nitrogen. During its use a film was deposited on the surface of the anode. Attempts to fully identify this material have not been successful, but energy dispersive analysis performed in the scanning electron microscope produced the spectra seen in Fig. 8. The spectrum from the unused wire is that of tungsten; the spectrum from the used wire shows additional peaks which corresponded to sulfur and chlorine K-radiation. It was found subsequently that the epoxy used in the construction of the counter contained both these elements.

Culhane et al report that epoxy cement used in construction of sealed counters caused the performance to eventually degrade and this was eliminated by using glass-to-metal seals [25].

Figure 9 shows that the resolution also shows a count rate dependency.

Bender and Rapperport reported one counter that exhibited no change in mean pulse height with count rate and another which showed no shift

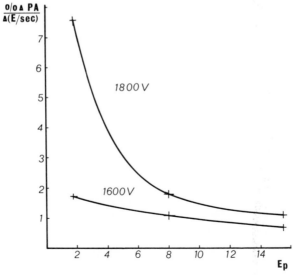

FIG. 6—*The percent change in pulse amplitude for a given change in rate of energy absorption is a function of photon energy (modified from Bender and Rapperport [19].)*

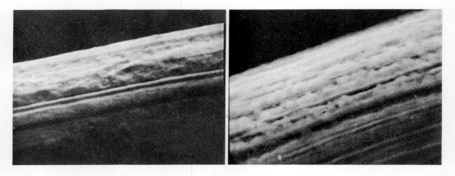

FIG. 7—*Surface of clean* (left) *and used proportional counter anode wire. Thin film containing chlorine and sulfur is deposited with use. Sealed counter, xenon and nitrogen filled.*

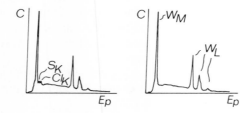

FIG. 8—*Total number of counts, (C) versus photon energy obtained by energy dispersion analysis from clean anode wire* (right) *and used anode wire in Fig. 7. Sulfur and chlorine peaks were not present in spectra from clean wire. Possible source is epoxy use in construction of counter.*

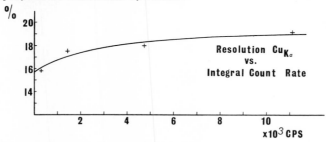

FIG. 9—*The relative variance (FWHM/$\bar{V}$) of the pulse height distribution is also affected by count rate changes. Counter gas, P-10.*

for low energy photons at low values of gas gain. Culhane et al describe counters which are less sensitive to count rate than most of those described by Bender and Rapperport [19,25]. The lack of consistency in the data supports Speilberg's conclusion that the decrease in pulse amplitude with count rate, especially at low counting rates, is not inherent in proportional counters. However, unless eliminated this characteristic could unduly complicate spectrum unfolding techniques.

Resolution

The variation in the output of proportional counters is due to several causes: (1) fluctuations in the initial ionization process, (2) fluctuations in avalanche production, (3) electronic noise, and (4) variation in design parameters of the counter either with time or with position in the detector.

Since these sources of variation are independent, the variances, σ^2, will sum in quadrature, hence

$$\sigma_V = \sqrt{\sigma_N{}^2 + \sigma_A{}^2 + \sigma_E{}^2 + \sigma_p{}^2}$$

where subscripts N, A, E, and P refer to the four sources of fluctuations listed above and V refers to the output pulse.

The fluctuations in the ionization process place a limit upon the resolution of the detectors. Several investigators have proposed statistical models for the prediction of the resolution of ionization devices [26-29]. One feature common to the results of these studies is that the ionization processes is not described by Poisson statistics and the fluctuations encountered are two to three times less than would be expected by assuming Poisson statistics.

The energy of the photon is partitioned between ionization and excitation of the atoms of the detecting material. The relative probabilities of the two events are fundamental in determining the inherent resolutions of the detector. Fano investigated this problem initially. From his work came the Fano factor, which is used in the prediction of energy resolution both in solid-state and gaseous ionization devices. By definition the Fano factor is the variance, $\sigma_N{}^2$, in the number of ions produced, $\overline{N}$.

$$F = \sigma_N{}^2/\overline{N}$$

The Fano factor is identical to the Lexis ratio used in other areas of statistics [30]. For a process which can be described by Poissonian statistics the Fano factor is one. The Fano factor applies to the initial ionization process only, not to the avalanche formation.

Because of the similarity in form of equations for the inherent resolution which one obtains by assuming Poisson statistics and those where one assumes a known Fano factor, the Fano factor is sometimes mistaken for a correction factor. It is, however, a characteristic of the material and its interaction with the ionizing radiation. Because of this, it can be predicted and has been manipulated to some degree [29,31]. If ionizations are the event being detected, increasing the probability of an ionizing event while reducing the probability of excitation will improve the resolution. Alkhozov, Komar, and Vorob'ev improved the resolution in a gaseous ionization chamber by using a gas mixture which increased the yield of ions. Acetylene, the ionization potential of which is less than the excitation energy of argon, was used to replace methane in an argon-polyatomic gas ionization chamber. The number of ions produced by the incident 5.68 MeV α-particle was increased and the resolution improved [31]. Similar approaches to the improvement of proportional counters for X-rays should be investigated.

If one knows the Fano factor and assumes $\overline{N}$ proportional to E_p,

$$E_p = \epsilon \overline{N}$$

$$\epsilon = E_p/\overline{N}$$

then ϵ is the average energy required per ion produced.

Since

$$\sigma_N{}^2/\overline{N} = F$$

then

$$\sigma_N/\overline{N} = \sqrt{F/\overline{N}}$$

and

$$\sigma_N/\overline{N} = \sqrt{F\epsilon/E_p}$$

For the initial ionization process, therefore, the fractional standard deviation decreases as E_p increases and increases as ϵ increases. Fano factor for solid-state devices and gaseous mixtures are not greatly different. F for argon is about 0.22, but F equal to 0.09 for argon-acetylene mixtures has been reported [31]. (F for Si(Li) detectors is about 0.15.) The superior inherent resolution of solid-state detectors is due to the lower value of ϵ rather than better Fano factors. Various values of ϵ for pure argon for X-rays excitation have been reported; 27 eV is a good approximate value, while 3.6 eV is the value used for silicon [32]. Alkhozov et al reported an $\epsilon = 20.5$ eV for A-C_2H_2 against 26.0 eV for A-CH_3 [31]. Assuming equal Fano factors, the fractional standard deviation due to ionization fluctuations ($\sigma_N/\overline{N}$) for a gaseous ionization device will be roughly 2.7 times larger than for a silicon device.

Although gas amplification is noiseless, fluctuations in the avalanche process do occur and contribute to the variance of the pulse height distribution. Campbell predicts that the variance due to the avalanche is nearly equal to that caused by the primary ionization process [28].

Since the variation due to the primary process is inherently larger in gas detectors and the variance due to the gas amplification must be added to this, the limits on the resolution of proportional counters are much poorer than those on solid-state detectors.

Campbell has derived an equation for proportional counter resolution which he feels is most optimistic [28]:

$$\sigma_V/\overline{V} = 0.11 \ E_p^{-1/2}$$

$\sigma_v/\overline{V}$ is the fractional standard deviation of the pulse due to monoenergetic radiation of energy E_p.

Mulvey and Campbell proposed a similar equation [27]:

$$\sigma_V/\overline{V} = 0.19 \ E^{-1/2}$$

which gives predictions of $\sigma_v/\overline{V}$ higher than experimental values. They suggest that the experimental data is limited by the relationship:

$$\sigma_V/\overline{V} = 0.15 \ E^{-1/2}$$

Figure 10 shows these relationships plotted along with curves showing what might be expected from solid-state detectors. At one time the most severe limitation of the Si(Li) detector appeared to be the electronic noise level. However, the last year or so has seen electronic noise levels reduced significantly. Noise levels of 100 eV have been demonstrated, and 50 to 60 eV can be expected.[4,5] At this time the limit appears to be set by the absorption of the window, noise generated by luminescence of the specimen, and the necessity of maintaining the surface of the silicon wafer free from contamination. If these problems can be overcome and 50 eV or less electronic noise levels are realized, the solid-state detector is preferable because of its inherently better resolution. If Campbell's prediction is assumed to be a limit for proportional counters and the curve for 50 eV equivalent electronic noise for Si(Li) detectors, it is apparent that the solid-state device will be superior. If 50-eV electronic noise becomes practical and the ultra-soft radiation can get to the sensitive volume of the detector, the Si(Li) detector can cover the range from boron through the periodic table. With this detector the K-radiations from equal amounts of carbon, nitrogen, oxygen, and fluorine will be resolved.

Williams and Sara discuss the dependence of proportional counter resolution on design variables [15]. The detector voltage was found to be the most critical followed by gas pressure, anode diameter, and cathode diameter in that order.

Conclusion

The role of the gas flow proportional counter in energy dispersive analysis depends upon two conditions: (a) the practicality of energy dispersive analysis in the range below 1 keV, and (b) the availabilty of a better detector in the near future.

---

[4] See p. 217.
[5] Gedcke, D., Ortec, Inc., Oak Ridge, Tenn., personal communication.

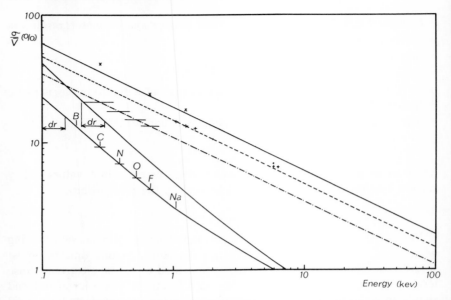

FIG. 10—*Comparison of relative standard deviation in pulse height distribution as predicted and measured for flow proportional counter and predicted for Si(Li) detectors. Solid-state detector curves contain both electronic noise and ionization fluctuations, flow proportional counter curves ionization fluctuations only. Data points X = Henke [20], • = Bisi and Zappa [26], + = Sutfin and Ogilvie. Straight lines represent following equations: —— $0.19E^{-1/2}$, —— $0.15E^{-1/2}$, — • — • — $0.11E^{-1/2}$ for the proportional counter resolution. Upper curve represents resolution of Si(Li) detector with 100 eV electronic noise, lower curve 50 eV electronic noise. Position of K-lines of light elements are shown on 50 eV Si(Li) curve. Horizontal bars represent full width at half-maximum (FWHM) of a symmetrical pulse distribution curve. (dr) represents range in which detection may be possible but signal to noise ratios will be unfavorable.*

Energy dispersion analysis below 1 keV, in general, will be quite complicated at best. Automatic data processing will be required for even qualitative analysis. Although the detection of ultra-soft X-rays with the gas flow proportional counter has been demonstrated, the poor resolution and lack of stability of the proportional counter will further complicate the analysis. One detector which can cover the whole range of energies rather than a tandem arrangement will simplify data processing.

If priorities are to be assigned to detector development for energy dispersion analysis, then the improvement of the solid-state detectors should receive major emphasis, especially interfacing the detector with the rest of the equipment so that unsupported thin films may be used as windows.

There are indications in the literature of approaches to the problems of improving the resolution and stability of proportional counters. These should not be neglected, because the proportional counter will continue to play an important role in wavelength dispersion analysis, in energy dispersion analysis in situations where the operating conditions for semiconductor detectors cannot be maintained, and at present it is the only detector available for analysis of ultra-soft X-rays.

References

[1] Saffir, A. J., Ogilvie, R. E., and Harris, R. S., *Proceedings of the Fourth National Conference on Electron Microprobe Analysis,* EPASA, Paper 53, 1969.

[2] Dolbey, R. M., in *X-Ray Optics and X-Ray Microanalysis,* Pattee, H. H., Coslett, V. E., and Engstrom, A., eds., Academic Press, New York, 1963, p. 483.

[3] Trombka, J. I. and Adler, I., in *Advances in Electronics and Electron Physics,* Supplement 6, Tousimis, A. J. and Marton, L., eds., Academic Press, New York, 1969, p. 313.

[4] Bearden, J. A., "X-ray Wavelengths and X-ray Atomic Energy Levels," NSRDS-NBS 14, National Bureau of Standards, 1967.

[5] Shimizu, R, Murata, K., and Shinoda, G., *Technology Reports of the Osaka University,* Vol. 16, No. 691, 1966, p. 131.

[6] Vogel, R. S. and Fergason, L., *Review of Scientific Instruments,* Vol. 37, No. 7, 1966, p. 934.

[7] Ranzetta, G. V. T. and Scott, V. D., *Journal of Scientific Instrumentation,* Vol. 44, 1967, p. 983.

[8] Baun, W. L., *Review of Scientific Instruments,* Vol. 40, No. 8, 1969, p. 1101.

[9] Hendee, C. F., Fine, S., and Brown, W. B., *Review of Scientific Instruments,* Vol. 27, 1956, p. 531.

[10] Sharpe, J., *Nuclear Radiation Detectors,* 2nd ed., Wiley, New York, 1964, p. 22.

[11] Sharpe, J., *Nuclear Radiation Detectors,* 2nd ed., Wiley, New York, 1964, pp. 63 and 70.

[12] Korff, S. A., *Electron and Nuclear Counters,* 2nd ed., Van Nostrand, New York, 1955, p. 59.

[13] Korff, S. A., *Electron and Nuclear Counters,* 2nd ed., Van Nostrand, New York, 1955, Chapter 4.

[14] Rose, M. E. and Korff, S. A., *Physical Review,* Vol. 59, 1941, p. 850.

[15] Diethorn, W., "A Methane Proportional Counter System for Natural Radiocarbon Measurements," Technical Report NYO-6628, Atomic Energy Commission, Oak Ridge, Tenn., 1956.

[16] Williams, A. and Sara, R. I., *International Journal of Applied Radiation and Isotopes,* Vol. 13, 1962, p. 229.

[17] Zastawny, A., *Journal Scientific Instrumentation,* Vol. 44, 1967, p. 395.

[18] Evans, R. D., *The Atomic Nucleus,* McGraw-Hill, New York, 1955, Chapter 26.

[19] Bender, S. L. and Rapperport, E. J., in *The Electron Microprobe,* McKinley, T. D., Heinrich, K. F. J., and Wittry, D. B., eds., Wiley, New York, 1966, p. 405.

[20] Henke, B. L. in *Advances in X-ray Analysis,* Muller, W. M., Millett, G., and Fay, M., eds., Plenum Press, New York, Vol. 7, 1964, p. 460.

[21] Henke, B. L., in *Advances in X-ray Analysis,* Muller, W. M., Mallet, G., and Fay, M., eds., Plenum Press, New York, Vol. 8, 1965, p. 269.

[22] Caruso, A. J. and Kim, H. H., *Review Scientific Instruments,* Vol. 39, 1968, p. 1059.

[23] Speilberg, N., *Review of Scientific Instruments,* Vol. 38, 1967, p. 291.

[24] Speilberg, N., *Third National Conference on Electron Microprobe Analysis,* EPASA, Paper 34, 1968.

[25] Culhane, J. L., Herring, J., Sanford, P. W., O'Shea, G., and Phillips, R. D., *Journal of Scientific Instruments,* Vol. 43, 1966, p. 908.

[26] Fano, U., *Physical Review*, Vol. 27, No. 1, 1947, p. 26.

[27] Mulvey, T. and Campbell, A. J., *British Journal of Applied Physics*, Vol. 9, 1958, p. 406.

[28] Campbell, A. J., *Norelco Reporter*, Vol. 14, 1967, p. 103.

[29] Van Roosbroek, W., *Physical Review*, Vol. 119, No. 5A, 1965, p. A1702.

[30] Evans, R. D., *The Atomic Nucleus*, McGraw-Hill, New York, 1955, Chapter 27.

[31] Alkhozov, G. D., Komar, A. P., and Vorob'ev, A. A., *Nuclear Instruments and Methods*, Vol. 48, 1967, p. 1.

[32] Dearneley, G. and Northrop, D. C., *Semiconductor Counters for Nuclear Radiations*, 2nd ed., Barnes and Noble, Inc., 1966, p. 72.

J. C. Russ[1]

Light Element Analysis Using the Semiconductor X-ray Energy Spectrometer with Electron Excitation

REFERENCE: Russ, J. S., "Light Element Analysis Using the Semiconductor X-ray Energy Spectrometer with Electron Excitation," *Energy Dispersion X-ray Analysis: X-ray and Electron Probe Analysis, ASTM STP 485*, American Society for Testing and Materials, 1971, pp. 217–231.

ABSTRACT: Limitations on light element analysis include the low fluorescent yield of low atomic number elements and the interference of closely spaced K-lines and the L- and M-lines of heavier elements. The paper discusses the relative contributions of electronic noise and statistical line width for low-energy X-rays and the effect of detector window thickness, along with the possibility of reducing or eliminating the window. Optimum accelerating voltages for light elements are calculated, and the importance of takeoff angle is stressed. Examples are given of the minimum detectable limits obtainable at the present state of the art; the possibilities for development in the near future are discussed.

KEY WORDS: semiconductor devices, X-ray spectrometers, X-ray fluorescence, background noise, spectrochemical analysis, detection, energy absorption, resolution

The rapidly expanding use of semiconductor X-ray energy spectrometers, especially with scanning electron microscopes and electron probe microanalyzers, is reflected by the other papers in this volume. Several of these papers have pointed out that an important (perhaps *the* important) shortcoming of these devices is their inability to analze elements lighter than about sodium ($Z = 11$, $K\alpha = 1.04$ keV). Historically it is interesting to note that this was also a barrier to the conventional crystal-diffraction spectrometer for many years, and that it has been overcome only relatively recently. The interest in light element detection and analysis is great, however, and in considering the future application of the semiconductor spectrometer to X-ray analysis it is important to examine the present state of the art for light element detection and make reasonable predictions of future expectations. (The paper by Sutfin and Ogilvie considers another approach to light element detection, using the gas flow proportional counter.)

[1] Applications consultant, Jeolco (USA), Inc., Medford, Mass. 02155. Personal member ASTM.

In this discussion it is necessary to distinguish between detection and analysis. The ability to detect soft X-rays is achieved much more easily, and in fact oxygen ($Z = 8$, $K\alpha = 523$ eV) X-rays from SiO_2 have been detected with a fairly convenitonal present-day system [1].[2] However, the count rate for the oxygen X-rays was so low that the analysis of oxygen, that is, determining how much is present in realistic specimens, is out of the question at present.

This paper will consider the factors that limit the ability to analyze soft X-rays with the semiconductor spectrometer. They include the low fluorescent yield of K-lines from low Z elements, potential interferences by L- and M-lines from other elements present in the specimen, the difficulty of resolving K-lines from adjacent light elements, and the absorption of low-energy X-rays by the entrance window of the spectrometer and in the specimen itself. Some of these same factors are discussed in the paper in this volume by Sutfin and Ogilvie.[3]

Fluorescent Yields

Figure 1 shows the fluorescent yield as a function of atomic number. This is the fraction of ionizations produced in the atoms of the specimen that decay producing an X-ray photon. The competing process is Auger electron production, in which the decay of one orbital electron to a lower energy level causes another electron to be ejected, leaving with surplus kinetic energy equal to the difference between its original energy level and the energy released by the decay.

The probability of X-ray photon emission becomes very low for light elements. The curve is described by $W = Z^4/a + Z^4$ where Z is the atomic number and a is a constant for each electron shell [2]: $a_K = 1.12 \times 10^6$; $a_L = 1.02 \times 10^8$. Figure 1 shows that W for carbon, for instance, is less than 0.5 percent, while for iron it is nearly 30 percent. Hence if we wish to analyze carbon in an ordinary 0.4 percent carbon steel, the ratio of X-rays from the iron and the carbon would be greater than 15,000 to 1. This neglects the different ionization cross sections of carbon and iron atoms and the effects of absorption in the specimen or the detector window, but the order of magnitude is illustrative.

This appears to be a formidable obstacle to light element analysis. However, the conventional electron probe microanalyzer has the same problem and still manages to perform acceptable analysis for elements down to beryllium. The minimum detectable limit is not so good for light elements as for somewhat heavier ones, but analysis of concentration levels in the range 0.1 to 1 percent is possible in many materials of practical interest.

---

[2] The italic numbers in brackets refer to the list of references appended to this paper.

[3] See p. 197.

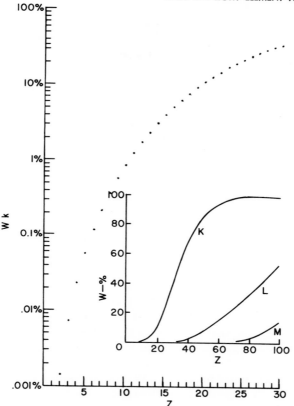

FIG. 1—*Fluorescent yield (W) as a function of atomic number (Z) averaged for each electron shell (inset) and detailed for low energy Kα X-rays from light elements.*

This suggests that if other limitations of the semiconductor X-ray spectrometer itself can be overcome the analysis of light elements should be possible.

K-L and K-M Interferences

If the *K*-lines of the light elements were all alone in the spectrum at low energies, their detection and analysis would be hard enough. But there are also the *L*- and *M*-lines of heavier elements. Figure 2 shows the location of just the major *L*- and *M*-lines. More complete data for higher energy lines are shown in the paper by Fitzgerald and Gantzel[4] in this volume. It must be remembered that the marks represent the locations of the peak centers; their breadth depends on the spectrometer resolution, which is controlled primarily by electronic noise at these low energies. At the present state of

[4] See p. 3.

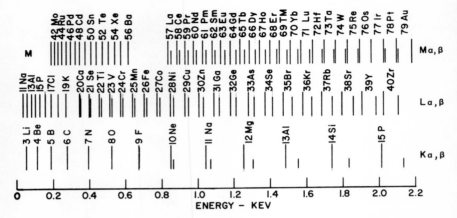

FIG. 2—*Principal low energy emission lines showing potential interelement interferences.*

the art this implies a full width at half maximum (FWHM) greater than 100 eV, which in the future might conceivably drop to 60 eV. This would make it possible to resolve the oxygen K from titanium, vanadium, or chromium L X-rays if those elements were present in the specimen.

These interferences can be overcome by curve fitting and stripping techniques in a computer. It seems likely that because of the other reasons for interest in attaching computers to X-ray energy spectrometers, many users will have this facility by the time the resolution improves and the other problems are overcome. Even so, the plethora of lines in the low energy range makes it imperative that the spectrometer resolution improve if the peaks are to be well enough defined and separated for practical analysis.

Resolution and Analysis

Considering only the K-lines, the analysis of small amounts of one element in the presence of large amounts of its nearest neighbor in the periodic table is a serious and practical test of performance. The FWHM describes the ability to resolve peaks of approximately equal height; even the full width at tenth maximum (FWTM), which has not yet been accepted as a specification by all manufacturers, does not adequately relate to concentrations of 1 percent or so.

To illustrate this problem a series of aluminum-magnesium alloys were prepared, ranging from 1 percent aluminum in 99 percent magnesium to 1 percent magnesium in 99 percent aluminum. Analysis was performed using a semiconductor X-ray energy spectrometer on a Jeolco JSM-U3 scanning electron microscope. The system resolution measured at 5.9 keV was 163 eV, and the noise width with a pulser 102 eV, making this representative of the current state of the art.

Figure 3 shows the spectra obtained. The peak heights of the aluminum and magnesium lines from the 50-50 alloy are not equal because of the difference in fluorescence yield, and the absorption of X-rays in the specimen itself and in the spectrometer window. The accelerating voltage used was 40 kV, the spectrometer was mounted with a 32-deg takeoff angle, and the entrance window was nominal 1 mil beryllium.

Considering first the aluminum peak, we see that it is resolved clearly right down to 1 percent, and using just the peak height we can compare count to concentration as shown below:

Al, %	Peak Count	Normalized Count
1.......	427	1.35
5.......	1665	5.27
10......	3262	10.32
50......	15826	50.09
90......	28464	90.10
95......	30038	95.08
99......	31276	99.01

The peak count was corrected by subtracting the constant white radiation background, but not for the interfering peak. The normalized count compares quite well with concentration but shows an error at low percentages due to the interfering magnesium peak. Integrated peak areas might be preferred to peak height, but the overlap makes the area difficult to determine.

For the magnesium peak the error is greater, and in fact the 1 percent magnesium peak is not resolved.

Mg, %	Peak Count	Normalized Count
1.......	...	...
5.......	1211	7.00
10......	1985	11.47
50......	8502	49.13
90......	15465	89.37
95......	16383	94.68
99......	17131	98.93

In this case the normalized count is lower than magnesium content because of absorption in the specimen, but is artificially raised at low percentages because of substantial interference from the aluminum peak. Examination of the curves in Fig. 3 shows that the low energy side of each peak has a sloping tail. This is due to X-rays ionizing some silicon atoms near the surface of the detector; the charge near the surface is hard to collect (a partially dead layer) and so the resulting pulse is lower in energy than it should be, producing a low energy tail on the peak.

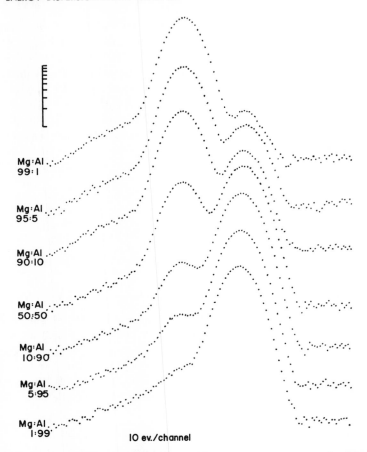

Mg:Al
99:1

Mg:Al
95:5

Mg:Al
90:10

Mg:Al
50:50

Mg:Al
10:90

Mg:Al
5:95

Mg:Al
1:99

10 ev./channel

FIG. 3—*X-ray spectra from magnesium-aluminum alloys measured with 163-eV FWHM (at 5.9 keV) resolution detector.* (Note: *The vertical scale is logarithmic.*)

This tailing is a serious problem that limits the ability to analyze light elements in the presence of heavier ones. The shapes of the tails are not uniform, and in fact their cause is not understood in detail.

To gain some insight into the effect of future improvements in resolution on the ability to analyze light elements, we repeated the above measurements with *poorer* resolution, obtained by changing the amplifier time constants from 10 to 2 μs. This increases the equivalent electronic noise (pulser resolution 173-eV FWHM) for a system resolution at 5.9 keV of 215 eV, still better than many systems in routine use at the present time. As shown in Fig. 4, the 1 percent aluminum peak and the 1, 5, and 10 percent magnesium peaks are now difficult or impossible to detect without elaborate curve fitting. (This may not be practical when tailing affects the shape of the peak.)

This comparison of 163 and 215-eV FWHM at 5.9 keV can be extended to other systems resolutions and their light element performance. The difference between 215 and 163-eV resolution at 5.9 keV makes an enormous difference in the ability to analyze aluminum ($K\alpha$ = 1.487 keV) and magnesium ($K\alpha$ = 1.254 keV). A further inprovement in FWHM resolution at 5.9 keV from 163 to 135 eV, for instance, would be accompanied by a further reduction in noise contribution (to 50-eV pulser resolution) so that an even greater improvement in resolution for light elements would be obtained. For 50-eV pulser resolution we would expect 77-eV FWHM for the magnesium peak, and it would then be possible to analyze down to 0.1 Mg in aluminum (provided tailing also is eliminated).

The line broadening resulting from the statistics of ionization in the detector is

$$E_{statistics} = 2.35\sqrt{E\epsilon F}$$

where:

E = X-ray energy,

ϵ = energy required for each ionization (3.8 eV in cooled silicon), and

F = Fano factor describing the degree of correlation in successive ionizations by an X-ray (0.13 in cooled silicon).

This broadening adds in quadrature to the electronic noise, expressed as equivalent energy and measured as the width of a pulser generated peak

$$\text{FWHM} = \sqrt{E^2_{statistics} + E^2_{noise}}$$

This means that a noise contribution of 50 eV would yield total peak resolutions for light elements as shown in Fig. 5. This would permit the separation of equal intensity K-lines down to beryllium, and resolution of a carbon peak one tenth as high as a neighboring nitrogen peak. Furthermore, this electronic noise would produce a low energy cutoff so low that boron (Z = 5, $K\alpha$ = 185 eV) would be completely resolved from noise, and beryllium (Z = 4, $K\alpha$ = 110 eV) would be partially resolved.

Such a level of light element performance would make the semiconductor X-ray energy analyzer very competitive with conventional crystal diffraction spectrometers. The improvement to 50-eV equivalent noise in the electronics seems likely, and the reduction if not elimination of low energy tailing also seems possible.

Window Absorption

Most present-day semiconductor X-ray energy analyzers are provided with entrance windows of 1 mil (0.025 mm) thick beryllium. This serves to maintain the vacuum isolation of the cryostat and to stop any light photons or electrons backscattered from the specimen, which would also ionize the detector. Beryllium is used because of its low absorption coefficient and high strength, and because it does not produce any X-rays that would be seen in the spectrum.

However, the absorption of low energy X-rays in beryllium is still severe. Figure 6 shows the percent transmission through 1 and 1/2 mil beryllium windows. Reducing the window thickness lowers the energy at which a useful percentage of the incident X-rays are transmitted so that one more element can be analyzed.

To illustrate this we have measured the X-ray intensity coming from specimens containing oxygen ($Z = 8$, $K\alpha = 523$ eV), fluorine ($Z = 9$, $K\alpha = 677$ eV), sodium ($Z = 11$, $K\alpha = 1.041$ keV), magnesium ($Z = 12$, $K\alpha = 1.254$ keV), aluminum ($Z = 13$, $K\alpha = 1.487$ keV), silicon ($Z = 14$, $K\alpha = 1.740$ keV), phosphorus ($Z = 15$, $K\alpha = 2.015$ keV), chlorine ($Z = 17$, $K\alpha = 2.622$ keV), and calcium ($Z = 20$, $K\alpha = 3.690$ keV) and used this intensity to calculate the minimum detectable limit.

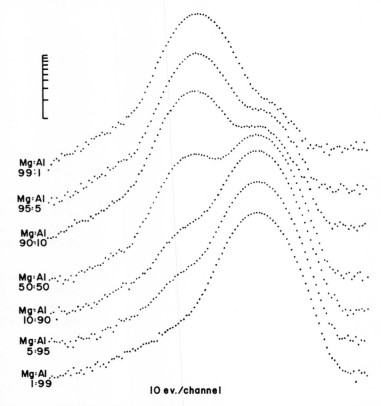

FIG. 4—*X-ray spectra from magnesium-aluminum alloys measured with 215-eV FWHM (at 5.9 keV) resolution detector. (Note: The vertical scale is logarithmic.)*

Optimum Accelerating Voltage

In order to optimize the sensitivity of the system, we chose the accelerating voltage for each element to maximize the count rate. This depends on the absorption of X-rays in the specimen, since as incident electron energy increases more X-rays are produced, but as they are produced deeper and deeper in the specimen, fewer are emitted from the surface. The result is a curve of count versus accelerating voltage that rises steeply to a maximum and then falls, described by

$$I = Kie^{-q\mu} \frac{\sin \phi}{\cos \psi} V^2 \left[\frac{V - V_0}{V_0} - \log \frac{V}{V_0} \frac{\alpha^2}{6V_0} (2V^3 - 3V^2V_0 + V_0^3) \right]$$

$$\alpha^2 = q\mu \left(\frac{\sin \phi}{\cos \psi} \right)$$

where:

I = characteristic X-ray intensity,
K = constant describing capture cross section,
V = accelerating voltage,
V_0 = absorption edge,
μ = mass absorption coefficient,
ϕ = angle of incident electron beam from normal,
ψ = angle of emitted X-ray beam from surface,
q = constant describing electron penetration into specimen, and
i = absorbed electron current.

Figure 7 shows the measured results. The X-ray takeoff angle was 32 deg and the maximum accelerating voltage available in the Jeolco JSM-U3 is 50 kV, so that for elements heavier than about chlorine optimum was not reached. The accelerating voltage used for each element is listed in a table later.

It should be noted that the optimum voltage will vary not only with instrument geometry but also specimen composition. If a mixture of elements is present, the absorption will be different and thus change the optimum voltage. And if the mixture is inhomogenous the situation will, of course, be further affected.

Minimum Detectable Limit

Statistically we require for a small peak to be detected that it must rise above background by twice the standard deviation of the background (noise and white radiation), or a minimum detectable concentration

$$C_{MDL} = \frac{2\sqrt{2} \sqrt{I_B}}{I_o \sqrt{t}} C_{standard}$$

where:

I_B = background count,

I_o = net count,

t = counting time, and

C_{Standard} is the concentration of the element in the standard.

The net count is integrated over 1.2 times the FWHM to maximize [4] the ratio of $I_o/\sqrt{I_B}$. A counting time of 1000 s was used in all cases. The table below lists the data and Fig. 8 shows the results graphically as a function of energy.

Element	K energy	Specimen	kV	I_o/I_B	½ Mil Window I_o^a	C_{MDL}	1 Mil Window I_o^a	C_{MDL}
O......	0.523	SiO_2	20	0.09	0.00025	9.9	...	...
F......	0.677	LiF	26	4.55	0.016	0.24	0.0027	8.15
Na.....	1.041	NaCl	33	24.3	0.124	0.022	0.055	0.078
Mg....	1.254	Mg	37	177.9	1.657	0.0052	1.225	0.0061
Al.....	1.487	Al	41	295.8	3.589	0.0027	2.662	0.0032
Si......	1.741	SiO_2	44	114.5	1.163	0.0036	0.938	0.0040
P......	2.015	$Ca_{10}F_2(PO_4)_6$	47	14.6	0.725	0.0051	0.476	0.0062
Cl.....	2.622	NaCl	50	105.8	2.585	0.0033	2.018	0.0037
Ca.....	3.690	$Ca_{10}F_2(PO_4)_6$	50	79.6	2.068	0.0027	1.952	0.0029

[a] Counts per second per picoampere of beam current.

Since several of the standard specimens used are not pure elements, the data are affected by X-ray absorption due to the other elements in the specimen. This is particularly evident in the results for phosphorus. Hence, the dashed line in Fig. 8 lies below the points for elements not measured on pure specimens. These data do not consider the absorption that may take place in a real specimen, so that in general the practical minimum detectable limit in a specimen of practical interest may be several times worse, but certainly elements from magnesium up (with a 1 mil window) and sodium up (with a 1/2 mil window) can be analyzed down to 0.1 percent in most situations, where serious interferences or absorptions are not present. This is comparable to conventional crystal diffraction spectrometers used on electron probe microanalyzers.

The counting time used could be reduced either by using higher X-ray fluxes produced by increasing the incident beam current or increasing the solid angle covered by the detector, which was only 0.00136 steradian. Different mechanical design easily could increase this value to about 0.007 steradian, a gain of five times, while increasing the takeoff angle from 32 to 40 to 45 deg. This would reduce the absorption path length in the specimen and significantly improve the sensitivity of the system to light elements. By increasing the detector size from 4 to 10 mm diameter (and

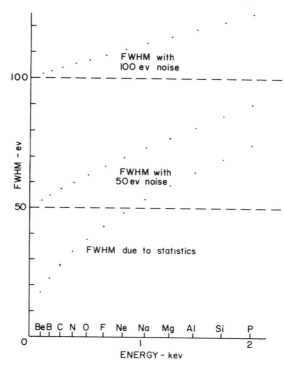

FIG. 5—*Combination of statistical and noise broadening for light element lines, assuming 50 and 100 eV equivalent electronic noise.*

accepting some loss of resolution) the solid angle could be made as large as 0.04 steradian.

Takeoff Angle

The takeoff angle of the X-ray detector is defined as the angle between the nominal specimen surface and the line of sight from the detector to the X-ray source. Since the X-rays are generated at a depth of up to 1 μm or more in the specimen (depending on the accelerating voltage used) the generated X-rays must travel a distance through the material of the specimen that is proportional to the cosecant of the takeoff angle.

The mass absorption coefficients for soft X-rays in most materials are quite high. For instance, for 1 keV X-rays in iron ($Z = 26$) the mass absorption coefficient is over 10^4, and 99.976 percent of the X-ray intensity is absorbed in 1 μm. Thus for light element analysis it is especially important to minimize this path length by maximizing the takeoff angle. The emission (Fig. 9) is proportional to $e^{-\csc \Phi}$. It is clearly impractical to use takeoff angles below 20 to 25 deg for light element analysis, and at least 40 deg would be preferred.

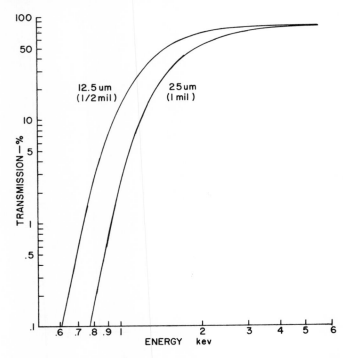

FIG. 6—*Absorption of low energy X-rays by beryllium window.*

Windowless Systems

Since scanning electron microscopes and electron probe microanalyzers operate with a vacuum anyway, it seems obvious to attempt to remove the beryllium window altogether. In practice, this is not so simple. As mentioned before, the window stops incident light photons and electrons and maintains vacuum isolation. The instruments are generally light tight, and very little light is emitted by the specimen under electron bombardment. Furthermore, it is possible, although not easy, to put a grid maintained at −50 kV in front of the detector to deflect backscattered electrons (this must be shielded with another grid at ground potential to prevent unwanted deflection of the incident electron beam or the emitted secondary electrons). Electromagnetic deflection of the backscattered electrons away from the X-ray detector is also possible.

But the semiconductor detector and its associated electronics are cooled to nearly liquid nitrogen temperature, and so act as a cold trap. Any oil, water vapor, or other contamination will collect there and degrade detector performance by causing electrical interference or acting as an absorbing window on the detector. The conventional oil diffusion pumped vacuum

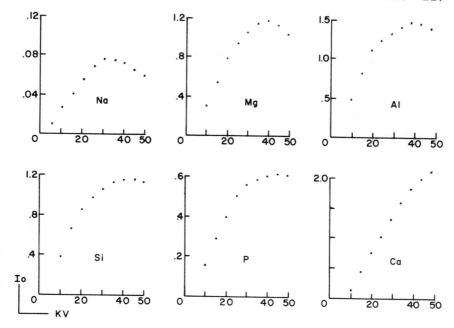

FIG. 7—*Variation of count rate (counts per second-picoampere) with accelerating voltage (kV).*

system used in commercial instruments is very satisfactory and practical, and it is unlikely that it will be abandoned in the near future. Furthermore even if it were, the outgassing of specimens and contaminants (even finger-prints) would still result in contamination buildup on the detector, which would be extremely difficult to remove.

For these reasons it seems desirable to maintain some vacuum isolation between the crystal of the X-ray detector and the electron beam instrument. The present 1 or 1/2 mil beryllium windows are strong enough to withstand atmospheric pressure, and thinner windows such as those used on gas counters cannot do so without rupturing or leaking. However, this difficulty could be overcome by using a window too thin to support atmospheric pressure but protected by an airlock opened only in the vacuum of the electron beam instrument. The window can then be extremely thin, and made of some material such as polypropylene or formvar, rather than beryllium. This possibility also is discussed in the paper by Aitken and Woo[5] in this volume.

Transmission curves for these materials show that as much as 70 percent of incident carbon radiation and 45 percent of incident nitrogen radiation can be transmitted. The presence of absorption peaks due to oxygen

[5] See p. 36.

and carbon in the window materials themselves produces an irregular transmission versus energy curve, but this is not a serious drawback. Further information on these windows, which are also used with gas flow proportional counters, is given in the paper by Sutfin & Ogilvie[3] in this volume.

These thin windows would not be adequate to stop high energy electrons, and so deflecting electrostatic or magnetic fields would be required. And the airlock, required to permit removal of the X-ray spectrometer or cleaning of the electron beam instrument (or specimen introduction if an airlock is not incorporated in the instrument) would have to be compact to avoid mechanical interference, and reliable to prevent accidental opening in air which could result in the total failure of the detector. But if these practical design problems can be overcome, while the noise resolution of the spectrometers continues to improve and peak tailing is reduced, then the prospects for using solid-state X-ray energy analyzers for the practical analysis of elements down to boron appear very promising.

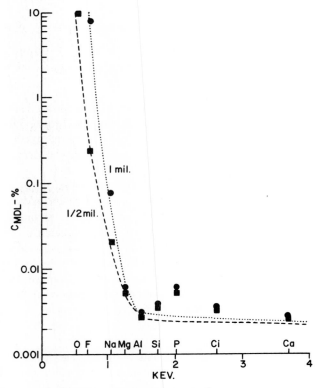

FIG. 8—*Calculated minimum detectable limit* (C_{MDL}) *of low-energy X-rays using 163-eV FWHM (at 5.9 keV) resolution detector with ½ and 1 mil beryllium windows.*

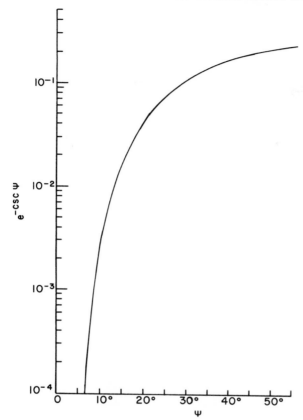

FIG. 9—*Variation of emitted intensity with takeoff angle* Ψ *showing effect of absorption path length through specimen.*

References

[*1*] Elad, E. et al, "The Use of an Energy Dispersive X-ray Analyzer in Scanning Electron Microscopy," *1969 Nuclear Science Symposium,* Institute of Electrical and Electronic Engineers.
[*2*] Burhop, *The Auger Effect,* Cambridge University Press, Cambridge, England, 1952.
[*3*] Rossland, *Philosophical Magazine,* Vol. 45, 1923, p. 65.
[*4*] Sutfin, L. and Ogilvie, R., Paper 54, *Proceedings of the 4th National Conference on Electron Microprobe Analysis,* 1969.

R. L. Myklebust[1] *and K. F. J. Heinrich*[1]

Rapid Quantitative Electron Probe Microanalysis with a Nondiffractive Detector System

REFERENCE: Myklebust, R. L. and Heinrich, K. F. J., **"Rapid Quantitative Electron Probe Microanalysis with a Nondiffractive Detector System,"** *Energy Dispersion X-ray Analysis: X-ray and Electron Probe Analysis, ASTM STP 485,* American Society for Testing and Materials, 1971, pp. 232–242.

ABSTRACT: Nondiffractive detector systems on electron probe microanalyzers have decreased the time required for a qualitative analysis from 30 or 40 min to 1 or 2 min. The nondiffractive technique may be extended to the quantitative determination of an element by comparing either the peak height or the area beneath the peak of the signal from the specimen with that from the pure chemical element or a standard of known composition run under identical conditions. Results are given for quantitative analyses of several materials with a lithium-drifted silicon detector, used for X-ray lines above 1 keV, and a gas flow proportional detector, for X-ray lines from 0.16 to 1.5 keV. Quantitative analyses can be accomplished only when the experimental conditions are carefully selected and maintained throughout the measurement. They are then sufficiently precise to warrant the use of a "matrix" correction procedure to obtain concentrations. An analog curve analyzer is useful for resolving overlapping peaks and for integrating the areas beneath the peaks.

KEY WORDS: electron probes, microanalysis, gas flow, proportional counters, multichannel analyzer, quantitative analysis, X-ray spectrometers, evaluation, tests, lithium, silicon, solid state counters, X-ray spectra, electron diffraction

The Li-drifted silicon detector [*1,2*][2] has become an increasingly useful tool for nondiffractive electron probe microanalysis. Its main advantage is the speed with which a complete qualitative analysis can be performed. A complete wavelength scan with diffracting spectrometers on an unknown specimen usually requires 30 to 40 min, while sufficient information can be frequently obtained by nondiffractive analysis in 1 to 2 min. For this

---

[1] Institute for Materials Research, National Bureau of Standards, Washington, D. C. 20234.

[2] The italic numbers in brackets refer to the list of references appended to this paper.

reason, the number of points investigated on a specimen can be substantially increased when the nondiffractive technique is employed.

Quantitative determinations of the elements observed in the spectra can be also performed with this same detector. This is done by comparing either the peak height or the area beneath the peak of a line emitted by the specimen with that emitted by a pure chemical element or a standard of known composition, run under identical conditions. The technique increases the number of elements that can be determined simultaneously. Previously this number was limited by the number of available X-ray spectrometers [2].

Instrumentation

The lithium-drifted silicon detector with a 0.05-mm beryllium window can be used to detect X-ray photons of energy above 1 keV. Within this range of energies are the K-radiations from atomic numbers 12 to 32 and the L-radiations from atomic numbers 30 to 83. In order to extend the capabilities of the nondiffractive system to softer X-rays, and hence to lower atomic numbers, a flow proportional detector with a thin nitrocellulose window was used for X-rays from 0.16 to 1.5 keV. The K-radiation of elements from atomic numbers 4 to 13 and the L-radiation from atomic numbers 23 to 35 are covered by this detector with little or no interference from M or higher lines.

A block diagram of the essential components of a solid-state detector system is shown in Fig. 1. More comprehensive descriptions of X-ray detection with solid-state devices can be found in the papers by Bowman

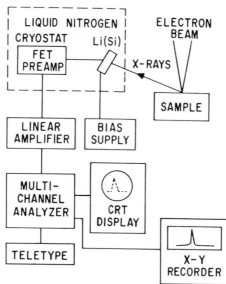

FIG. 1—*Block diagram of a solid-state detector system.*

et al [1] and Fitzgerald et al [2]. Several companies now manufacture these detectors, which can be mounted directly on an electron probe micro-analyzer, as described in the latter reference.

The components of a gas flow proportional detector system are basically the same as for the solid-state detector. Discussions of nondiffractive X-ray analysis using flow proportional detectors have been given by Dolby [3] and by Birks and Batt [4]. The detector used in the current work was operated with 90 percent argon–10 percent methane at atmospheric pressure. This detector is identical to the flow proportional detector used in our electron probe on the spectrometer for soft X-radiation (5 to 100 Å).

The energy resolution of a spectrometer system usually is defined as the width of the measured X-ray line at half the height of the peak above background. A good resolution is desirable, not only because of the problem arising from overlapping lines of different elements, but also in view of the relation between resolution and the ratio of the net signal and the background (line-to-background ratio). In qualitative analysis, a multi-channel analyzer is used to sort out the signal pulses according to their height. The contents of the analyzer memory can be displayed on a cathode ray tube, recorded by means of an *x-y* recorder, or transferred to a digital computer.

The relatively poor resolution of the nondiffractive spectrometer frequently produces partial overlap of the observed analytical lines. In such situations, deconvolution techniques (spectrum stripping) are very helpful. There exist many digital computer programs which accomplish this task. However, the digital procedure is time consuming unless provisions exist for rapid transfer of the contents of the analyzer memory into the computer. A rapid viable alternative is the graphic output of the spectrum by means of an *x-y* recorder, with subsequent analysis by an analog curve resolver.

For digital analysis in spectrum stripping, it is desirable that each peak should extend over at least 6 to 8 channels. If we assume that each peak has a width of 400 eV, and if our wavelength region of interest ranges from 1 to 10 keV, it follows that the minimum number of channels required is $(8 \times 9000)/400 = 180$. However, the spectra are easier to interpret visually if the number of channels is larger. We use 400 channels for the analysis of a spectrum, reserving another 400 channels for the storage of a standard spectrum which can be displayed on the analyzer oscilloscope simultaneously with the spectrum of the specimen.

For the performance of quantitative, or even qualitative analyses, the experimental conditions—detector bias, amplifier gains, pulse-shaping times, and multichannel analyzer setttings—must be carefully selected and maintained without change throughout the measurements. The conditions should be chosen in a manner such that the entire energy range of interest (1 to 10 keV or more for the lithium-drifted silicon detector) is displayed by the multichannel analyzer. The channel positions of two X-ray lines of widely different energies, such as germanium $K\alpha$ (9.886 keV) and aluminum $K\alpha$ (1.487 keV) are then determined. The channel numbers of these lines are plotted versus energy. Since the channel number is a linear function of energy, the energy of an X-ray line can be determined directly from the channel number. This calibration remains valid for several months except when the X-ray counting rate is too high. The point at which the shift of the spectrum at high counting rates becomes significant may depend on the characteristics of both the detector system and the multichannel analyzer, and must be determined for each system. In our instrument, the spectra shift toward higher energy when the multichannel analyzer dead-time meter reads higher than 10 percent.

Quantitative Analysis

A quantitative analysis of any specimen requires knowing all the elements it contains. Therefore, as a preliminary examination, a complete qualitative analysis should be made. A typical series of qualitative analyses on a steel specimen (Fig. 2) shows iron in most areas, as well as manga-

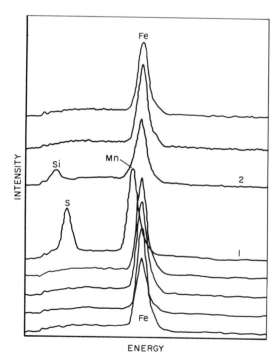

FIG. 2—*Qualitative analysis of steel. Spectrum 1 is of a manganese sulfide inclusion. Spectrum 2 was obtained from a slag inclusion containing silicon, iron, and manganese.*

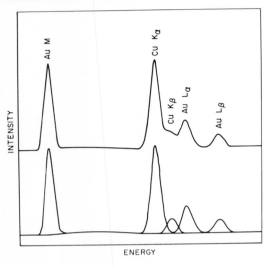

FIG. 3—*40Cu-60Au alloy. The upper spectrum is the original plot. The peaks are shown in the lower spectrum resolved by means of an analog spectrum analyzer.*

nese and sulfur in a sulfide inclusion; the analysis of an oxidic area in the same specimen shows the presence of iron, silicon, and manganese. These spectra required one minute each and were obtained with a lithium-drifted silicon detector which had a resolution of 500 eV. The peaks of adjacent elements in the iron group were not resolved with this detector; however, overlapping peaks can be separated either by spectrum stripping or by improving the detector. In another case, the spectrum from a copper-gold alloy (Fig. 3), in which the copper $K\beta$-line was not resolved completely, was examined with the analog curve analyzer so that all peaks could be resolved. A more recently available silicon detector, which has a resolution between 200 and 300 eV, has been used also to examine the same copper-gold alloy. The spectrum as produced by this detector, was fully resolved since the resolution is sufficient to completely separate iron $K\alpha$ (6.404 keV) from iron $K\beta$ (7.058 keV).

The quantitative analysis procedure is described using as examples two series of standard alloys prepared at the National Bureau of Standards. The spectra from the four gold-copper alloys (SRM 482) (Fig. 4) and similar spectra from the four gold-silver alloys (SRM 481) were obtained with the 500-eV resolution detector. The compositions were computed both by comparing the peak heights above background on the alloys with those on the pure metals, and by comparing the integrated areas beneath these peaks on the alloys with those on the pure metals. The peak heights and backgrounds were measured directly from the *x-y* chart recordings. However, the number of counts in the peak channel on the multichannel analyzer would serve equally well.) The backgrounds for each standard and alloy were determined by interpolating the continuum through the peaks. The areas beneath the peaks were integrated with the analog curve analyzer. The intensity ratios (k-values) were then calculated by the equation:

$$k = I^*_A/I_A \times C_A$$

where:

I^*_A = intensity above background of a characteristic X-ray line of Element A emitted by the specimen,

I_A = intensity above background of the same line emitted by a standard for Element A, and

C_A = concentration of Element A in the standard.

The k calculated always refers to the pure element.

Since X-rays are generated and absorbed in the same manner as for diffractive analysis, the usual data correction procedures ("matrix corrections") must be applied. The computer program, COR [6], which was used to calculate concentrations, corrects the intensity ratio for absorption, atomic number, characteristic fluorescence, and continuum fluorescence. Table 1 lists the resulting k-values and calculated values obtained for the concentrations together with the certified concentrations of the gold-copper alloys while Table 2 lists the same data for the gold-silver alloys. The estimated standard deviations obtained from three independent measurements of the same spectra from the copper-gold alloys are listed in Table 3. This readout error is larger for gold than for copper since the gold Lα-peak is much smaller than the copper Kα-peak (Fig. 4). The

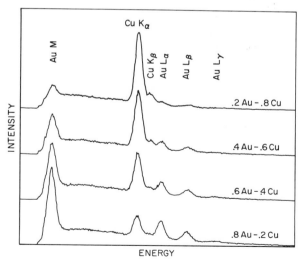

FIG. 4—*Spectra from NBS standard gold-copper alloys (SRM 482).*

TABLE 1—*Copper-gold analyses, NBS SRM 482.*

| | C_E | | k | |
C_S	Peak Height	Integrated	Peak Height	Integrated
		Copper—$K\alpha$		
		Standard: Pure Cu		
0.1983	0.211	0.216	0.245	0.250
0.3964	0.411	0.408	0.452	0.450
0.5992	0.607	0.616	0.639	0.650
0.7985	0.797	0.807	0.813	0.825
		Gold—$L\alpha$		
		Standard: Pure Au		
0.8015	0.784	0.797	0.725	0.741
0.6036	0.626	0.631	0.550	0.556
0.4010	0.422	0.443	0.350	0.370
0.2012	0.221	0.256	0.175	0.204

NOTE—C_S = certified standard concentration.
C_E = experimental concentration.
k = intensity ratio (specimen/standard).

gold $M\alpha$-peak might also have been used for this analysis since the peak-to-background ratio is better than for the gold $L\alpha$-peak; however, the corrective procedures for M-lines are not as well known as for L-lines. The values obtained for copper are slightly poorer than normally would be expected from microprobe analysis; however, they are sufficiently precise

TABLE 2—*Silver-gold analyses, NBS SRM 481.*

| | C_E | | k | |
C_S	Peak Height	Integrated	Peak Height	Integrated
		Silver—$L\alpha + L\beta$		
		Standard: Pure Ag		
0.1996	0.180	0.201	0.143	0.160
0.3992	0.383	0.392	0.317	0.325
0.5993	0.584	0.610	0.508	0.535
0.7758	0.770	0.771	0.708	0.710
		Gold—$L\alpha$		
		Standard: Pure Au		
0.8005	0.846	0.785	0.825	0.758
0.6005	0.611	0.612	0.575	0.576
0.4003	0.409	0.413	0.375	0.379
0.2243	0.237	0.236	0.213	0.212

NOTE—C_S = certified standard concentration.
C_E = experimental concentration.
k = intensity ratio (specimen/standard).

to warrant the use of a correction procedure. The results for gold are expected to improve with longer counting times. Further experiments are in progress.

TABLE 3—*Peak height measurement errors in copper-gold analyses.*

	Copper			Gold	
Certified C	Measured C	Estimated Standard Deviation	Certified C	Measured[a] C	Estimated Standard Deviation
		15 kV Excitation			
0.799	0.792	0.004	0.201	0.200	0.004
0.599	0.578	0.004	0.401	0.441	0.017
0.396	0.362	0.001	0.604	0.635	0.011
0.198	0.191	0.001	0.802	0.833	0.002
		20 kV Excitation			
0.799	0.779	0.003	0.201	0.185	0.024
0.599	0.581	0.002	0.401	0.396	0.015
0.396	0.376	0.001	0.604	0.582	0.011
0.198	0.213	0.004	0.802	0.778	0.027

[a] At 20 kV, the gold $L\alpha$-line was measured on the same scale as the copper $K\alpha$-line. At 15 kV, the scale used for the gold $L\alpha$-line was expanded to improve the measurement.

When an overlap of peaks could not be resolved by the detector, the analog curve analyzer served for spectrum stripping as well as for integrating the areas beneath the peaks. Synthetic spectra have been created to test the validity of the procedure by summing spectra from pure elements. The spectrum from a 40Cr-60V alloy (Fig. 5) was produced by the multichannel analyzer by integrating for 40 percent of the total time (0.8 min) on pure chromium and 60 percent (1.2 min) on pure vanadium. It can be observed that the chromium $K\alpha$-(5.415 keV)-peak is apparently larger than expected because of the contribution from the vanadium $K\beta$-line (5.427 keV). Since the vanadium $K\alpha$-peak in the "alloy" was 60 percent of the peak in pure vanadium, the $K\beta$-peak in the "alloy" also should be 60 percent of the $K\beta$-peak in pure vanadium. Once this ratio has been determined, the spectrum could be resolved (Fig. 5). (For an analogous procedure on titanium and vanadium see Ref 7.) The results of four analyses made on the synthetic chromium-vanadium alloys are shown in Table 4. No "matrix" corrections were needed for these data since the spectra were generated from pure elements.

A nondiffractive detector system is particularly valuable for the identification of minerals. If the material can be identified semiquantitatively, it may be unnecessary to perform a precise quantitative analysis of the elements present. We have collected spectra from a large number of known

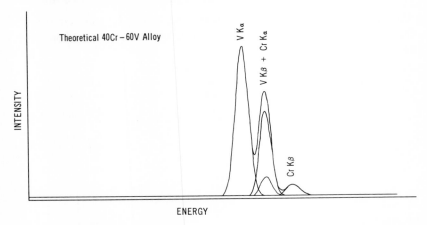

FIG. 5—*Resolved and unresolved spectra of a synthetic chromium-vanadium "alloy" obtained with the 280-eV resolution lithium-drifted silicon detector.*

TABLE 4—*Vanadium-chromium analyses, synthetic alloys.*
Standards: Pure elements

Vanadium—$K\alpha$		Chromium—$K\alpha$	
C	C_E	C	C_E
0.8	0.830	0.2	0.19
0.6	0.615	0.4	0.39
0.4	0.405	0.6	0.60
0.2	0.200	0.8	0.82

NOTE—C = calculated concentration.
C_E = integrated experimental concentration.

minerals with both the high resolution silicon detector and with the flow proportional detector. Table 5 shows the results of a quantitative analysis for aluminum, sodium, and oxygen of Dawsonite in an oil shale. The

TABLE 5—*Dawsonite analysis, $NaAlCO_3(OH)_2$.*
Standards: Al_2O_3 and NaCl

Element	C	C_E		k	
		Peak Height	Integrated	Peak Height	Integrated
Al........	0.19	0.24	0.22	0.18	0.16
Na........	0.16	0.13	0.11	0.074	0.065
O.........	0.56	0.51	0.57	0.30	0.36
Total......	0.91	0.88	0.90		

NOTE—C = calculated concentration.
C_E = experimental concentration.
k = intensity ratio (specimen/standard, multiplied by concentration of element in standard).

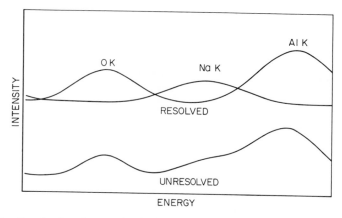

FIG. 6—*Resolved and unresolved spectra of a Dawsonite specimen obtained with a gas flow proportional detector.*

spectrum from the flow proportional detector (Fig. 6) is broad and needs to be resolved before measuring the peaks. However, the results are acceptably accurate. A more precise measurement, in fact, may not improve the results for elements of atomic number below ten, since the uncertanties in the absorption correction [8] set a limit to the accuracy of the analytical procedure.

The relatively high dead-time losses in the multichannel analyzer are an inherent disadvantage of the use of such analyzers for quantitative purposes. This effect can be circumvented by using a single-channel analyzer for each element to be measured. In view of the high stability of the silicon detector, it is not necessary to open the single channel to the entire pulse height distribution of the peak. Thus, in cases of partial line overlap, the resulting interference can be minimized by judicious setting of the threshold of the pulse height analyzer. This technique was used in the analysis of lunar specimens for the elements; aluminum, silicon, titanium, and iron [9]. However, it must be borne in mind that the low line-to-background ratios obtained nondiffractively preclude the accurate analysis at low concentration levels (< 1 percent), unless very careful background measurements can be performed.

References

[1] Bowman, H. R., Hyde, E. K., Thompson, S. G., and Jared, R. C., *Science*, Vol. 151, No. 3710, Feb. 1966, pp. 562-568.
[2] Fitzgerald, R., Keil, K., and Heinrich, K. F. J., *Science*, Vol. 159, No. 3814, Feb. 1968, pp. 528-530.
[3] Dolby, R. M., *Proceedings of the Physical Society*, Vol. 73, No. 81, 1959.
[4] Birks, L. S. and Batt, A. P., *Analytical Chemistry*, Vol. 35, No. 778, 1963.
[5] Heinrich, K. F. J., "Quantitative Electron Probe Microanalysis," *NBS Special Publication 298*, National Bureau of Standards, 1968.

[6] Hénoc, J., Heinrich, K. F. J., and Myklebust, R. L., "COR, A Program for Computer Calculation for Quantitative Electron Probe Microanalysis," to be published.

[7] Snetsinger, K. G., Bunch, T. E., and Keil, K., *American Mineralogist,* Vol. 53, 1968, pp. 1771-1774.

[8] Yakowitz, H. and Heinrich, K. F. J., *Mikrochimica Acta,* Vol. 1, 1968, p. 182.

[9] French, D. M., Walters, L. S., and Heinrich, K. F. J., "Quantitative Mineralogy of an Apollo 11 Lunar Sample," *Geochimica et Cosmochimica Acta,* to be published.

J. R. Rhodes[1]

Design and Application of X-ray Emission Analyzers Using Radioisotope X-ray or Gamma Ray Sources

REFERENCE: Rhodes, J. R., **"Design and Application of X-ray Emission Analyzers Using Radioisotope X-ray or Gamma Ray Sources,"** *Energy Dispersion X-ray Analysis: X-ray and Electron Probe Analysis, ASTM STP 485,* American Society for Testing and Materials, 1971, pp. 243–285.

ABSTRACT: This paper reviews the present state of development and application of X-ray emission spectrometry using radioisotope X-ray and gamma ray sources for excitation, and energy discrimination and dispersion techniques for spectrum analysis.

The design and use of instrumentation is characterized by the choice of one of the two main methods of energy selection now employed. The balanced filter method is simple but restricted to sequential determinations of single elements and so finds its main use in small, hand-portable, battery-operated analyzers for field, laboratory, or industrial use. Energy dispersion with high resolution silicon or germanium detectors permits simultaneous collection of the whole excited spectrum and so promises rapid multielement analysis. Available equipment is more suited to laboratory conditions.

Relevant components of radioisotope X-ray spectrometers are described. Emphasis is placed on source-specimen-detector configurations, sources, source-target assemblies, absorption edge filters, and relevant properties of detectors. Equations for estimating the feasibility of a given analysis are derived, and interferences due to absorption, enhancement, and heterogeneity effects are discussed.

A major portion of the paper is devoted to a review of published applications and covers elemental analysis from aluminum to uranium, and measurement of coating thickness. References are made where appropriate to available portable, industrial, and laboratory instruments.

KEY WORDS: X-ray analysis, X-rays, gamma rays, dispersing, spectrum analysis, radioisotopes, X-ray fluorescence, source-target assemblies, X-ray filters, detector resolution, electronic systems, source-specimen-detector configurations, central source geometry, side-source geometry, annular source geometry, collimated beam geometry, feasibility equations, thin specimen analysis, interelement effects, particle size effects, classified applications, industrial applications, mining applications, laboratory applications, coating thickness determination, portable analyzers, multielement analyzers, on-stream analyzers, borehole logging probes, evaluations, tests

[1] Manager, Applied Research Division, Columbia Scientific Industries, Corp., Austin, Tex. 78702.

The first review [*109*][2] of radioisotope X-ray spectrometry contained 47 references and covered the period 1946, when the excitation of characteristic X-rays using a radioisotope source was first reported, to 1966. By that time several measurement techniques had been developed and many analyses proved feasible. Also a range of sealed, radioisotope, low-energy X-ray and gamma ray sources, and a few instruments, were commercially available.

Since then, over 120 papers have been published including 23 reviews and two book chapters [*20,117*]. Four international meetings have been held, devoting significant portions of their programs to the subject, including one contemporaneous with this paper.[3] Table 1 classifies the review

TABLE 1—*Classification of review articles.*

Author	Date	Reference	Subject Matter
Rhodes..........	1966	*109*	radioisotope X-ray spectrometry
Ansell...........	1967	*3*	sources
Berry...........	1967	*6*	mining applications
Brunner.........	1967	*16*	mining and industrial applications, with bibliography (in German)
Florkowski.......	1967	*46*	mining applications (in German)
Heath...........	1967	*55*	solid-state detectors
Kato............	1967	*67*	research and applications in Japan
Martinelli.......	1967	*90*	applications in France (in French)
Martinelli.......	1967	*91*	applications in France
Ostrowski........	1967	*102*	applications in Poland
Watt...........	1967	*134*	on-stream analysis in Australia
Alekseyev.......	1968	*2*	applications in mining (in Russian)
Bowie..........	1968	*13*	portable analyzers
Campbell........	1968	*20*	general
Campbell........	1968	*21*	X-ray absorption and emission
Dziunikowski....	1968	*39*	ore analysis (in Polish)
Kuusi...........	1968	*72*	applications in metallurgy
Aitken	1969	*1*	solid-state detectors
Birks...........	1969	*8*	current trends in X-ray fluorescence spectrometry
Clayton.........	1969	*28*	applications in geological assay, mining and mineral processing
Gallagher........	1969	*51*	applications in mining
Rhodes..........	1969	*117*	techniques in analysis and coating thickness measurement
Watt...........	1969	*135*	applications in Australian mineral industry

articles. They reflect the countries most active in the field and the specific work of each group.

The reason for this enormous increase in activity is the rapid worldwide application of the techniques developed in the early 1960's. These techniques are based on the detection and low resolution energy selection of

[2] The italic numbers in brackets refer to the list of references appended to this paper.
[3] *Third Symposium on Low Energy X- and Gamma-Ray Sources and Applications,* Boston College, Boston, Mass., June 1970 (not reviewed here).

characteristic X-rays using ordinary scintillation or proportional counters, and the use of filters for discriminating between X-rays from adjacent elements. The applications are mainly to field and industrial analyses, thereby exploiting the portability, low cost, and reliability of radioisotope sources. One type of instrument, the portable analyzer, is now available in a fairly standardized design from several manufacturers and is finding wide use, particularly in the mining industry. Of the 136 references listed, 36 describe specific applications of portable analyzers. Four reviews [6,13,28,51] are confined largely to the same subject.

An event of more general significance is the recent availability of lithium-drifted silicon and germanium detectors (Si(Li) and Ge(Li) detectors, respectively), whose energy dispersion is good enough to resolve K X-rays from adjacent elements down to sodium. This technological advance heralds a new era in X-ray emission spectrometry. So far, however, very few publications describe actual analyses with developed instruments. Since these detectors have to be cooled cryogenically and since multichannel analysis is required to properly exploit their resolution capabilities, it is expected that instruments using Si(Li) and Ge(Li) detectors will be applied first in laboratories.

In spite of the fundamental similarities between conventional and radioisotope X-ray spectrometery, it would be a mistake to present the radioisotope technique as an extension of conventional X-ray fluorescence analysis. It is already clear that new areas of application exist where X-ray fluorescence analysis has never before been considered. Instruments are being operated by people quite unfamiliar with conventional X-ray spectrometry and in environments hitherto foreign to the technique. For these reasons an attempt is made to present radioisotope X-ray spectrometry as an analysis technique in its own right.

The main aims of this paper are to describe relevant components, techniques, and instrumentation, and to review applications since 1966. Publications prior to 1966 are included only when they are either important original papers, or when a specific measurement first performed then has not been re-evaluated since.

Apparatus

A radioisotope X-ray fluorescence analyzer consists of the following basic components: a sealed radioisotope excitation source; a detection system which selects the energies of the characteristic X-rays excited and measures their intensities; and an electronic amplification and readout system whose output can be correlated with the elements present in the specimen and with their concentrations.

Figure 1 shows typical schematics of a portable, single element analyzer using filters for energy selection, and a more sophisticated multielement analyzer using a Si(Li) detector.

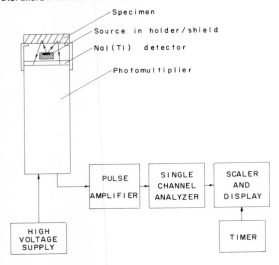

(a) Portable single element analyzer.

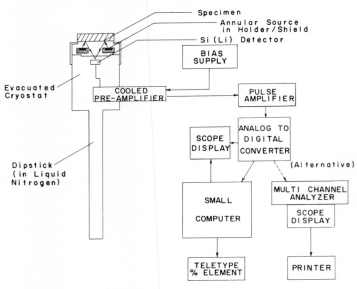

(b) Multielement analyzer.

FIG. 1—*Schematics of radioisotope X-ray fluorescence analyzers.*

The design and utilization of the basic components are reviewed in this section with special emphasis on geometry, sources, and filters.

Geometry

Typical source-specimen-detector configurations are shown in Fig. 2. In contrast to conventional X-ray spectrometry, dispersion is not produced by

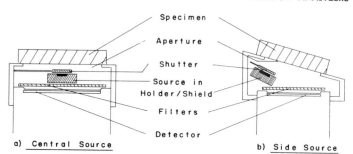

Specimen

Aperture

Shutter

Source in
Holder/Shield

Filters

Detector

a) Central Source

b) Side Source

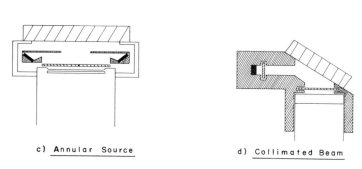

c) Annular Source

d) Collimated Beam

FIG. 2—*Typical source-specimen-detector arrangements.*

diffraction. Thus diffracting crystals, goniometers, and special collimators are not necessary, and overall geometrical efficiencies as high as 0.1 to 10 percent are normal. This means that quite adequate detector count rates (in the range 10^3 to 10^5 counts per second) are obtained from pure element specimens using sources whose emission is only 10^7 photons per second. This is some six orders of magnitude less than a typical X-ray tube output and is the reason why radioisotope sources can be used at all. It should be noted that the X-ray dose rate is also six orders of magnitude less than that from an X-ray tube.

A practical source-specimen-detector arrangement for radioisotope X-ray spectrometry has the following requirements:

1. Maximum geometrical efficiency.[4]

2. Negligible background due to direct transmission of radiation from source to detector.

3. Minimum background due to scatter or fluorescence from the instrument structure, especially the aperture that defines the measured specimen area.

4. Maximum fluorescence to scattered radiation intensity ratios from specimens of standard size; that is, 1 to 1.5 in. in diameter.

[4] This is defined as the photon intensity reaching the detector divided by that emitted from the source, assuming the specimen to be a perfect diffuse reflector.

5. Minimum sensitivity to movement of the specimen toward or away from the detector.

6. Enough shielding to reduce the radiation dose rate at accessible points outside the measuring head to below regulation levels.

Optimization of any one of the configurations shown in Fig. 2 is usually a compromise between several factors, and there is a great deal of scope for individual designs. "Central source geometry" is the most efficient for large detector windows and has found widest application with scintillation and proportional counters. The geometrical efficiency for a 2-in.-diameter detector window, a 2-in.-diameter specimen, and a ½-in.-diameter source approaches ten percent.

One of the characteristics of all the broad-beam arrangements is a so-called count rate distance plateau. Such a curve is shown in Fig. 3. The point of maximum count rate should be far enough from the source-detector assembly to be accessible to the specimen surface. This maximum can be shifted to larger distances by increasing the ratio of source holder to source diameter. This reduces the signal, especially if the diameter of the specimen becomes smaller than about one-and-a-half times the source holder diameter [108,117]. With the specimen set on a typical count rate distance plateau the count rate should decrease by less than one percent for a change in specimen distance of about 1 mm. Added advantages are

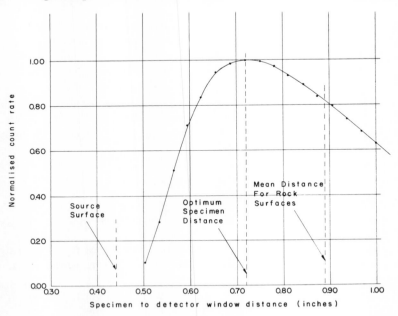

FIG. 3—*Typical count-rate distance relationship for central and side source geometries.*

relatively low sensitivity to specimen curvature and to surface roughness, although Darnley [33] has pointed out that a better position for measuring rough rock surfaces is on the linearly decreasing portion of the curve where the responses from high and low points on the specimen average out.

The position of the maximum of the count rate distance curve is a function not only of the instrument and source dimensions, which can be standardized, but also of the source energy. For source energies greater than about 20 keV, effective radiation penetration into the specimen becomes significant compared with the specimen to detector distance. The result is that the maximum on the curve moves towards the source with increasing source energy. The extent of this movement is a few millimeters, depending on the specimen density, for energies up to 100 keV.

"Side source" geometry is a useful alternative to the central source arrangement, particularly if the source and detector window diameters are comparable. Bednarek [5] and Ostrowski [5,102] show that a surprisingly long distance plateau can be obtained by moving the source away from the detector axis. Tolmie [128] and Duftschmid [36] suggest the use of side source geometry in portable analyzers, and we have developed it for use in a commercially available instrument. This has resulted in significantly better signals from standard-sized specimens than reported for central source geometry, even with a small probe (1.5 instead of 2 in. diameter).

Annular source arrangements are desirable for the small Si(Li) and Ge(Li) detectors but are used also with scintillation counters, particularly when the low specific activity of the source necessitates a large source area.

An annular ^{3}H/Zr source optimized for a 1.5-in.-diameter scintillation counter window (see Fig. 2c) was found to give 5 times the count rate from a given specimen using a source having 13 times the output of a central source [108]. Inclined annular source arrays have been used in preference to planar rings [31,33,53,108,124,137], but we have found that it is difficult to obtain a good count rate-distance characteristic for Si(Li) detectors with other than flat annular sources.

Uniform sensitivity across the specimen is a desirable feature, particularly when analyzing heterogeneous material. Figure 4 shows some typical plots of detector response against specimen position obtained with pin-head-sized specimens moved across the aperture. Relevant dimensions of the apparatus used to obtain the curves are given in Table 2. As is seen, central source geometry gives the worst response, with a deep minimum at the center of the field. Annular source geometry can give an essentially flat response [33].

Collimated beam geometry has little to recommend it unless the source emits penetrating radiation, when relatively heavy shielding is needed. Enomoto [44] optimized this arrangement for excitation of K X-rays of heavy elements using fairly high-energy gamma ray sources such as ^{137}Cs,

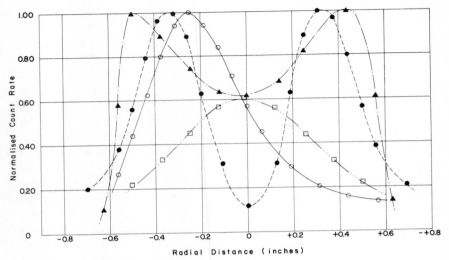

FIG. 4—*Detector response profiles across specimen surface.*

TABLE 2—*Relevant dimensions of source-target-detector configurations used to obtain data in Fig. 4.*

Configuration	Detector Diameter, in.	Source Holder Diameter, in.	Source Diameter, in.
Central source...............	1.5	0.7	0.4
Side source.................	1.2	0.6	0.4
Annular source.............	0.24	1.75 (OD)	1.4 (OD)
		0.75 (ID)	1.1 (ID)

and compared it with a broad beam geometry using ^{153}Gd. When excitation energies greater than about 100 keV are used, restriction of the angular divergence of incident and emitted radiation is important, since it limits the Compton scattered X-rays to a relatively narrow energy band which can be more easily discriminated against.

Sources

Alpha, beta, gamma, and X-ray emitting isotopes have all been used as sources to excite characteristic X-rays [109].

The most widely used excitation method is by primary or secondary gamma or X-ray sources, preferably those that emit one or a few spectral lines rather than a high proportion of continuous radiation. The major advantage of the preferred sources is that they excite characteristic X-rays efficiently with minimum background in the energy region of interest. A very wide range of such sources is now available in the desired energy range, namely, 5 to 150 keV, and with suitable specific activities, half-lives and encapsulations.

Bremsstrahlung emitters are secondary sources excited by beta particles and usually take the form of a mixture of beta emitter and target element. Bremsstrahlung spectra are continuous and have similar shapes to spectra emitted by X-ray tubes. Their main disadvantage is the continuous scattered background which accompanies excitation of characteristic X-rays. Early work [109] was concentrated on development of these sources and led to their use prior to the application of preferred X-ray emitting isotopes now available. Other secondary sources that are finding increasing use are gamma ray and X-ray excited X-ray sources.

Direct excitation by beta particles has been studied [109] but has two major disadvantages, excitation of bremsstrahlung background in the specimen and backscatter of a large proportion of the incident beta particles.

Excitation by alpha emitters becomes efficient below about 2 keV and complements X-ray and gamma ray excitation [120,123].

Primary X-ray and Gamma Sources—The criteria of a good source are: a simple line spectrum at an appropriate energy below 150 keV; no high-energy beta or gamma radiation; half-life of at least one year; specific activity high enough to yield a source emission of 10^7 to 10^8 photons/s from a 1 cm^2 surface; price not exceeding about two hundred U.S. dollars per 10^7 photons/s emission. Table 3 lists relevant properties of sources in use.

We have seen in the previous section that the overall dimensions of the source encapsulation are important in the design of instrument measuring heads. Due to self-absorption problems in this energy range it is quite difficult to make a compact source which combines efficient emission of the required radiation with effective containment of the radioactivity. Ansell [3] reviews this and discusses the design of disk, annular, point, line, and "conical" sources.

Figure 5 shows two standard encapsulations suitable for disk sources emitting energies in the range 10 to over 100 keV. A similar all-beryllium encapsulation is available for ^{55}Fe. The radioactivity is contained in the primary capsule by electrodeposition, or as a compacted powder briquette with aluminum, or as a ceramic enamel. Figure 6 shows an encapsulation for a "conical source." This was designed primarily to hold ^{241}Am for use in source-target assemblies. Conical sources also have been made containing ^{147}Pm, ^{109}Cd, or ^{57}Co.

Secondary Sources—Relevant properties of available ^{3}H and ^{147}Pm bremsstrahlung sources are listed in Table 4. In addition to these, Preuss [105] has compiled a catalog of characteristic X-ray and bremsstrahlung

TABLE 3—*Relevant properties of primary X-ray and gamma ray sources.*

Isotope	Half-Life, Years	Mode of Decay	Photon Emission			Normal Activity, mCi
			Energy (keV) and Type	Production,[a] %	Emission,[b] %	
^{55}Fe	2.7	electron capture	5.9, MnK X-rays	28.5	10 to 15	20
^{238}Pu	86.4	alpha	12 to 17, UL X-rays	10	5 to 10	30
^{75}Se	0.33	electron capture	10.5, AsK X-rays	40		
			140, gamma rays	54		
			270, gamma rays	56		
^{210}Pb	22	beta	11 to 13, BiL X-rays	24	10 to 15	10
			47, gamma rays	4		
			10 to 1000, bremsstrahlung	~2		
^{109}Cd	1.3	electron capture	22.2, AgK X-rays	107	80	1
			88.2, gamma rays [48]	4		
^{125}I^{c}	0.16	electron capture	27, TeK X-rays	138	100	1 to 10
			35, gamma rays	7		
^{241}Am	458	alpha	59.6, gamma rays	36	30	1 to 20
			14 to 21, NpL X-rays	37	0 to 20	
^{153}Gd	0.65	electron capture	103, gamma rays	20	? 20	1
			97, gamma rays	30	? 30	
			70, gamma rays	2.6	? 2	
			41, EuK X-rays	110	? 50	
^{57}Co	0.74	electron capture	136, gamma rays	8.8		
			122, gamma rays	88.9	? 80	1
			14, gamma rays	8.2		
^{137}Cs	30	beta	6.4, FeK X-rays	51	0 to 10	
			662, gamma rays	82	~ 80	

[a] Theoretical yield, percent photons per disintegration.
[b] Approximate practical output, percent photons per disintegration.
[c] Also the use of ^{125m}Te, with similar characteristics to ^{125}I, has been reported [126].

TABLE 4—*Revelant properties of available beta-excited sources*

| Isotope | Half-Life, Years | Max Beta Energy, keV | Target | Bremsstrahlung Emission | | Normal Activity, Ci |
				Efficiency Photons/Beta	Useful Energy Range, keV	
^{3}H.......	12.3	18	Ti	1.3×10^{-4}	2 to 10, plus TiK X-rays, 4.5	5
^{3}H.......	12.3	18	Zr	4×10^{-5}	2 to 10, plus ZrL X-rays, 2	3
^{147}Pm.....	2.6	220	Al	2×10^{-3}	10 to 70	0.5

NOTE—Special beta-excited sources reported as used are ^{35}S/Ba [75,76] and ^{14}C with various targets [130,131].

spectra obtained from various combinations of beta particle emitters and isotopes.

Source-target assemblies [3,28,53,107,109,133] consisting of an X-ray or gamma ray emitter and a nonradioactive target have found significant use particularly in "on-stream" analyzers [27,43,107,109,134]. By choice of any suitable target a wide range of spectrally pure characteristic X-rays can be obtained. The assembly has to be carefully designed to prevent too great a loss in photon output through geometrical inefficiency. Figure 7 shows a source-target assembly incorporated into a scintillation counter

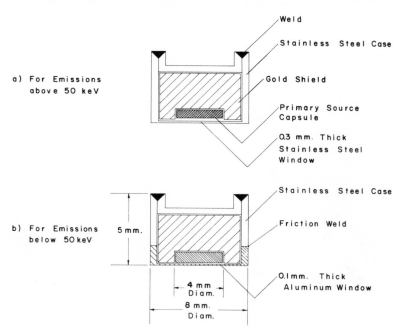

a) For Emissions above 50 keV

Weld
Stainless Steel Case
Gold Shield
Primary Source Capsule
0.3 mm. Thick Stainless Steel Window

b) For Emissions below 50 keV

5 mm.

Stainless Steel Case
Friction Weld
0.1 mm. Thick Aluminum Window

4 mm. Diam.
8 mm. Diam.

FIG. 5—*Typical source encapsulations.*

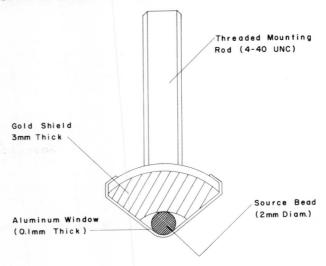

FIG. 6—*"Conical" source.*

[*107,109*]. This design has the merits of high geometrical efficiency and easy interchanging of targets. Targets can be made by mixing the powdered element with epoxy and casting in an appropriate mold.

An integral source-target assembly design is shown in Fig. 8a [*28*]. This has the advantage of being quite compact, but target changing is not as easy.

Giauque [*53*] has proposed the design of an annular version for use with solid-state detectors (see Fig. 8b), and discusses some preliminary measurements using one.

The spectral purity[5] and output of target characteristic X-rays from source-target assemblies has been studied [*107,109,133*] for the conical design described above, and the purity is likely to be similar for the other designs. Using a [241]Am primary source, the spectral purity for targets of atomic numbers above 45 (Rh) is greater than 90 percent but falls off rapidly for target X-ray energies less than about one third the primary energy. The output for the conical design is about 1×10^7 photons/s (normalized to 4π steradians) for a tin target and a standard 14 mCi [241]Am conical source.

Unwanted Radiation Emission—The detection of unwanted background due to high-energy radiation should no longer be a problem, as a result of progress in the design of sources and better availability of new and radiochemically pure isotopes. However, the capability of Si(Li) and Ge(Li) spectrometers to perform high resolution, high sensitivity analyses places more stringent demands upon the choice and purity of materials used in

[5] Defined as the percentage of target characteristic X-rays emitted.

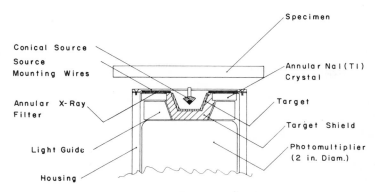

FIG. 7—*Source-target-detector assembly for scintillation counter.*

the source assembly. For example, gold K X-rays excited in the integral shield of ^{57}Co and ^{153}Gd sources penetrate the source capsule and can appear in the emitted spectrum [*18*].

Radiological Safety—Only radiation levels at the specimen position need be considered from the radiological safety point of view. Radiation emitted in other directions is well shielded because of detector background requirements, and levels are in the μR/h range. Typical X-ray dose rates at a distance of one foot from unshielded sources are in the 1 to 10 mR/h

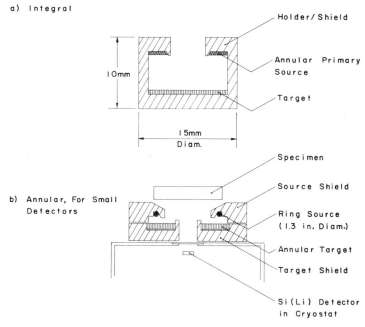

FIG. 8—*Other source-target assemblies.*

range. Dose rates at smaller distances are correspondingly higher since the sources can be assumed to be point emitters. Thus even with unshielded sources, exposure doses of 100 mR per 40-h working week are unlikely to be approached. Nevertheless, in routine operation, it is good practice to reduce the dose rate at the nearest accessible point outside the measuring head to as low a value as is conveniently possible. This is accomplished easily by incorporating a source shutter which is closed when a specimen is not in position. All commercially available portable analyzers have an automatic shutter of this type.

The second important feature of radiological safety is containment of the isotope. As discussed above, considerable effort has been put into encapsulating the sources so that even if they were smashed it would be very difficult for all the radioactivity to escape.

Thirdly, the proper licensing and operational procedures must be observed while acquiring, using, and disposing of radioactive sources. These are of local concern, and both the authorities and local service companies are usually willing to give every assistance to new radioisotope users.

Energy Selection

The resolving power[6] required of the energy selection apparatus varies from virtually nothing for certain very simple measurements to several hundred when complex groups of K and L X-rays have to be resolved.

In a surprisingly large number of applications, the substance to be analyzed has such a simple X-ray spectrum that the correct choice of source, together with a single filter on the detector window, is all the energy selection needed. Use of essentially monochromatic sources, especially source-target assemblies, can be made for preferentially exciting the required radiation with quite high efficiency. Carr-Brion [25] describes an elegant method of isolating a selected characteristic X-ray line and simultaneously minimizing matrix effects using two targets emitting close X-ray energies. One excites the required fluorescence while the second does not.

Wide use of the relatively coarse energy dispersion of scintillation and proportional counters also is made, mainly to discriminate against unwanted scattered radiation.

It is interesting to note that out of the 116 analyses classified in Table 5, 38 use only these very simple methods of energy selection. Of the rest, 60 use balanced filters and 18, high resolution Si(Li) or Ge(Li) spectrometers.

Figure 9 compares the resolving power of scintillation, proportional and Si(Li) spectrometers and crystal monochromators, and the discriminating power of balanced filters, with the separation between K X-rays from elements of adjacent atomic number [17]. The minimum requirement of a

---

[6] Defined as $E/\Delta E$, where E is the photon energy and ΔE the full width at half maximum (fwhm) of the peak.

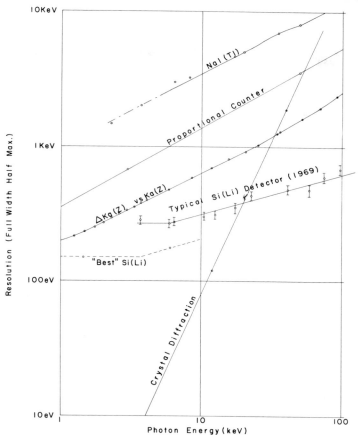

FIG. 9—*Energy resolution of various detectors compared with separation between* K *X-rays from adjacent elements.*

general purpose X-ray fluorescence analyzer is to be able to discriminate between *K* X-rays from adjacent elements. Balanced filters can do this for most elements in the periodic table by virtue of the alternation of the energies of *K* X-rays and *K* absorption edges.[7] However, the pass band containing the *Kα* X-ray of the element to be determined can also contain *Kβ*, *Lα*, and *Lβ* X-rays from other elements in the specimen. In spite of this and other drawbacks, balanced filters are the simplest and most widely used method of energy selection in radioisotope X-ray fluorescence analysis. The curves of Fig. 9 also show that Si(Li) spectrometers are now capable of resolving adjacent *Kα* X-rays down to 1 keV. At higher energies they exceed this performance considerably and are able to resolve

[7] There are no *K* absorption edges between the following pairs of *K* X-rays: Ti-V, Br-Kr, Ag-Cd, Pr-Nd, Ta-W, and U-Np.

TABLE 5—*Analyses.*

Element	Concentration Range	Material	Source and Activity	X-ray Excited	Filters	Detector	Instrument	Application	Detection Limit *(or 1σ precision)	Remarks	References
Aluminum	2 to 4% Al$_2$O$_3$	cement raw mix	^{3}H/Zr, 3Ci and ^{210}Po, 10 mCi	AlK	Al	proportional	on-line analyzer (fused pellets)	kiln-feed control	0.09% Al$_2$O$_3$	also see under Ca, Fe, and Si	28,129
Antimony	0.2 to 20%	antimony ores	^{35}S/Ba	SbK	Ag/Cd	scintillation	borehole logging probe	shallow hole and mine wall assay	0.2 to 0.3% Sb*		2,75,76 92
Argon	0.2 to 40%	air	^{55}Fe	AK	...	proportional	laboratory	feasibility study	0.5% relative*		62,102
Barium	0.01 to 10 mg	cellulose	^{241}Am, 3 Ci	BaK	...	Ge (Li), fwhm 600 eV	laboratory	sensitivity test	20 μg	counting time 650 min	31
	1 to 15%	barytes ores	^{241}Am-Sm and ^{241}Am-Ba, 14 mCi	BaK	...	scintillation	on-stream slurry analyzer	barytes flotation pilot plant control	0.01%	balanced target method used	25,27
Bromine	10 ppm to 5%	solutions	^{125}I, 5 mCi	BrK	...	Si(Li), fwhm 350 eV	laboratory	sensitivity test	...	matrix effects studied; also see under Mo, Rb, and Sr	64,65, 66
Cadmium	0.001 to 1%	solutions	^{241}Am, 10 mCi	CdK	...	Si (Li), fwhm 1.2 keV	laboratory	feasibility study	0.001%	also see under Cu, Mo, and Sn	59
Calcium	40 to 50%	cement raw mix	^{55}Fe, 2 mCi	CaK	...	scintillation	portable analyzer	feasibility study	0.2% CaO*		113
	1 to 5%	silicate rocks	^{55}Fe, 10 mCi	CaK	K/Ca	Si avalanche diode	portable analyzer	soil and rock type identification	0.5% CaO*	also see under K	63,80
	15 to 19% CaO	sinter mix	^{55}Fe, 7 mCi	CaK	...	proportional	on-line analyzer (pressed pellets)	blast furnace control	0.15% CaO*	also see under Si and Fe	109,136
	75 to 85% CaCO$_3$	cement raw mix	^{55}Fe, 2 mCi	CaK	...	proportional	on-stream slurry analyzer	kiln-feed control	0.35% CaCO$_3$*	...	28,109, 127
	40 to 50% CaO	cement raw mix	^{3}H/Zr, 3 Ci and ^{210}Po, 10 mCi	CaK	...	proportional	on-line analyzer (fused pellets)	kiln-feed control	0.16% CaO*	also see under Al, Fe, and Si	28,129
	40 to 50% CaO	cement raw mix	^{3}H/Ti, 5 Ci	CaK	...	proportional	on-stream powder analyzer	grinder control	0.4% CaO*		91,109

Element	Concentration	Sample	Source	Line	Ratio	Detector	Analyzer	Application	Precision	Notes	Ref
	40 to 45% CaO	cement raw mix	³H/Ti, 5 Ci	CaK	...	proportional	on-stream powder analyzer	kiln-feed control	0.2% CaO*	also see under Fe	10
	42 to 52% CaO	cement raw mix	¹⁰⁹Cd, 1 mCi	CaK	...	proportional	powder sample analyzer	feasibility study	0.17% CaO*	reduced particle size effects	86,102
Chromium	1 to 20%	steels	⁵⁵Fe, 7 mCi and ³H/Zr, 2.5 Ci	CrK	Ti/V	scintillation	portable analyzer	alloy analysis	0.1% Cr*		108
	0.5 to 8%	steels	²³⁸Pu, 20 mCi	CrK	Ti/V	scintillation	portable analyzer	alloy analysis	0.15% Cr*		113
	0.5 to 5%	electroplating bath solutions	³H/Zr, 4 Ci	CrK	Ti/V	scintillation	portable analyzer	alloy analysis feasibility study	0.17% Cr		114
Cobalt	0.001 to 2%	hydrocarbons	¹⁴⁷Pm/Al, 100 mCi	CoK	Mn/Fe	proportional	laboratory	routine laboratory analysis	0.001%		106
Copper	0.1 to 15%	copper ore pulps and rockfaces	³H/Zr, 12 Ci	CuK	Co/Ni	scintillation	portable analyzer	field assay	0.2% Cu*		13,33,34
	0.4 to 4%	copper ore pulps	³H/Zr	CuK	Co/Ni	scintillation	portable analyzer	draw control and sublevel caving	0.1% Cu*		29
	0.03 to 2.5%	copper ore pulps	¹⁰⁹Cd, 1 mCi	CuK	Co/Ni	scintillation	portable analyzer	feasibility study	0.03% Cu		113
	0.1 to 5%	copper ore pulps	³H/Zr, 2.5 Ci	CuK	Cr/Mn, Ni/Co	scintillation	portable analyzer	feasibility study	0.2% Cu*	special method for eliminating matrix effects	41
	0.1 to 10%	core samples	²³⁸Pu, 30 mCi	CuK	Co/Ni	scintillation	portable analyzer	core analysis	0.1% Cu*		100
	0.4 to 1.6%	manganese nodule pulps	²³⁸Pu, 20 mCi	CuK	Co/Ni	scintillation	portable analyzer	shipboard assay	0.05% Cu*	also see under Ni, Fe, and Mn	115
	0.1 to 2%	copper ores	²³⁸Pu	CuK	Co/Ni	scintillation	borehole logging probe	mine development	0.1% Cu*		28
	60 to 95%	copper alloys	¹⁰⁹Cd, 2 mCi	CuK	Co/Ni	scintillation	portable analyzer	alloy sorting	0.45% Cu*		108
	0.1 to 1%	solutions	²⁴¹Am, 10 mCi	CuK	...	Si(Li), 1.2 keV fwhm	laboratory	feasibility study	0.1% Cu	also see under Cd, Mo, and Sn	59
	0.1 to 26%	various Cu-Pb-Zn ore products	²³⁸Pu-Ga, 30 mCi	CuK	Cu	scintillation	on-stream slurry analyzer	mill control	0.08 to 1%*	also see under Pb, Sn, and Zn	43
Gold	0.01 to 1%	simulated ore samples	⁵⁷Co, 1 mCi	AuK	Hf/W	scintillation	portable analyzer	feasibility study	0.03% Au		114

TABLE 5—*Analyses* (continued).

Element	Concentration Range	Material	Source and Activity	X-ray Excited	Filters	Detector	Instrument	Application	Detection Limit *(or 1σ precision)	Remarks	References
Iron	20 to 70%	iron ores (pulps and cores)	^{238}Pu, 30 mCi	FeK	Cr/Mn	scintillation	portable analyzer	field assay, mine control	0.5% Fe*		49,50, 102
	1 to 10%; 2 to 12%	lead-zinc ores; manganese nodule pulps	^{238}Pu, 20 mCi	FeK	Cr/Mn	scintillation	portable analyzer	shipboard assay	0.24% Fe*		115
	5 to 50%	iron ores (crushed rock)	^{238}Pu, 10 mCi	FeK	Cr/Mn	scintillation	portable analyzer	strip mine control	~1% Fe*	X-ray backscatter also considered	37,38, 40,102
	25 to 45%	sinter mix	^{3}H/Zr, 4 Ci	FeK	Cr/Mn	proportional	on-line analyzer (pressed pellets)	blast furnace control	0.2% Fe*	also see under Ca and Si	109,136
	0.5 to 3%	cement raw mix	^{3}H/Zr, 3 Ci and ^{210}Po, 10 mCi	FeK	...	proportional	on-line analyzer (fused pellets)	kiln-feed control	0.05% Fe	also see under Ca, Fe, and Si	28,129
	0.5 to 3%	cement raw mix	^{3}H/Ti, 5 Ci	FeK	Cr/Mn	proportional	on-stream powder analyzer	kiln-feed control	0.003% Fe	also see under Ca	10
	0.1 to 5%	sand (slurry)	^{238}Pu, 10 mCi	FeK	Cr/Mn	proportional	on-stream slurry analyzer	pilot plant			28
	0.1%	lubricating oil	^{3}H/Zr, 2 Ci	FeK	Fe	G-M tube	routine assay	wear monitoring	0.01% Fe		102
Krypton	1 to 40%	air	^{109}Cd	KrK	...	proportional	...	...	0.5% relative*	...	62,102
Lead	0.05 to 30%	ore pulps	^{238}Pu, 30 mCi	PbL	Ga/Ge	scintillation	portable analyzer	field analysis	0.08% Pb	matrix and particle size studied	101
		lead ores	^{109}Cd	PbL	...	...	...	...			54
	1 to 8%	core samples	^{238}Pu, 30 mCi	PbL	Ga/Ge	scintillation	portable analyzer	core analysis	2% Pb*		100
		lead ores and mine walls	^{57}Co	PbK	...	scintillation	portable analyzer	mine control	0.25% Pb		79
		lead ores and core samples	^{75}Se	PbK	...	scintillation	portable analyzer	field assay	...	used nomograms	97,98, 125
	0.05 to 3%	leaded brass and steels	^{153}Gd, 0.6 mCi	PbK	W/Ir	scintillation	portable analyzer	alloy sorting	0.05% Pb		108
	0.1 to 1%	lead ores	^{57}Co	PbK	Re/Ir	scintillation	borehole logging probe	mine control	0.1% Pb		28

Element	Range	Sample	Source	Line	Filter	Detector	Application	Use	Sensitivity	Remarks	Ref
	0.1 to 32%	ores, zinc, concentrates, tailings	^{241}Am, 5 mCi, ^{57}Co, 1.2 mCi	...	...	scintillation	on-stream slurry analyzer	mill control	0.04 to 34% Pb*	absorption edge analysis; also see under Cu, Sn, and Zn	42,43
Manganese	0.5 to 5%	solutions and slurries	^{153}Gd, 1 mCi, ^{137}Cs, 1 Ci	PbK	...	scintillation	on-stream analyzer	feasibility study	0.4% Pb	comparison of sources also see under Fe, Cu, and Ni	44
	20 to 40%	manganese nodule pulps	^{238}Pu, 20 mCi	MnK	V/Cr	scintillation	portable analyzer	shipboard assay	0.75% Mn*		115
	0.2 to 1.5%	steels	^{3}H/Zr, 2.4 Ci	MnK	V/Cr	scintillation	portable analyzer	alloy sorting	0.14% Mn*		108
	0.5 to 5%	plating bath solutions	^{238}Pu, 20 mCi	MnK	V/Cr	scintillation	portable analyzer	feasibility study	0.04% Mn		114
Molybdenum	0.01 to 1% 1 to 20%	ore pulps copper concentrates	^{109}Cd, 2 mCi	MoK	Y/Zr	scintillation	portable analyzer	mine control	0.01% Mo		49,50
	0.001 to 0.3%	ore pulps	^{109}Cd, 1 mCi	MoK	Y/Zr	scintillation	portable analyzer	feasibility study	0.003% Mo		114
	40 to 55%	ore concentrates	^{147}Pm/Sm, 160 mCi	MoK	...	scintillation	laboratory	mine control	2% relative		9
	0.06 to 9%	steels	^{241}Am, 10 mCi	MoK	...	Si(Li), fwhm 1.2 keV	laboratory	feasibility study	0.01% Mo	also see under Cd, Cu, and Sn	59
	0.01 to 5%	steels	^{109}Cd, 2 mCi	MoK	Y/Zr	scintillation	portable analyzer	alloy sorting	0.04% Mo		108
	0.1 to 10%	steels	^{109}Cd, 1 mCi	MoK	Y/Zr	scintillation	portable analyzer	alloy analysis	0.1% Mo*		113
	0.01 to 1%	steels	^{125}I, 5 mCi	MoK	...	Si(Li), fwhm 350 eV	laboratory	sensitivity tests	<0.01% Mo		65
	2 ppm to 0.05%	solutions	^{125}I, 5 mCi	MoK	...	Si(Li) fwhm 350 eV	laboratory	sensitivity tests		matrix effects studied; also see under Br, Rb, and Sr; up to 3000-min count time	64,66
	10 ng to 1 mg	thin films	^{125}I, 5 mCi	MoK	...	Si(Li), fwhm 512 eV	laboratory	feasibility study	35 ng Mo		137
	0.002 to 0.5%	mine samples, copper concentrates, slags	^{125}I, 2 mCi	MoK	...	Si(Li), fwhm 640 eV	laboratory and on-stream analyzers	feasibility study	0.002 to 0.01% Mo		74
	0.01 to 7% MoO$_3$	ore slurries	^{147}Pm/Al-Ag 1 Ci	MoK	Y/Zr	scintillation	on-stream analyzer	pilot plant control	0.01% Mo	also see under Nb and Sn	27

TABLE 5—*Analyses* (continued).

Element	Concentration Range	Material	Source and Activity	X-ray Excited	Filters	Detector	Instrument	Application	Detection Limit *(or 1σ precision)	Remarks	References
Nickel	0.8 to 1.6%	manganese nodule pulps	^{238}Pu, 20 mCi	NiK	Fe/Co	scintillation	portable analyzer	shipboard assay	0.05% Ni*	also see under Fe, Cu, and Mn	*115*
	3 to 30%	copper alloys	^{3}H/Zr, 2.5 Ci	NiK	Fe/Co	scintillation	portable analyzer	alloy sorting	0.3% Ni		*108*
	0.1 μg to 1 mg	thin films	^{125}I, 10 mCi	NiK	...	Si(Li), fwhm 512 eV	laboratory	feasibility study	1 μg Ni	1000-min count time	*137*
Niobium	0.01 to 5%	ore slurries	^{147}Pm/Al-Ag, 1 Ci	NbK	Sr/Y	scintillation	on-stream analyzer	pilot plant control	0.01% Nb	also see under Mo and Sn	*27*
	0.001 to 1%	steels	^{125}I, 5 mCi	NbK	...	Si(Li), fwhm 350 eV	laboratory	sensitivity tests	~0.001% Nb		*65*
	0.06 to 2%	steels	^{109}Cd, 2 mCi	NbK	Sr/Y	scintillation	portable analyzer	alloy analysis	0.06% Nb		*108*
Potassium	1 to 5%	silicate rocks	^{55}Fe, 10 mCi	KK	Cl/K	Si avalanche diode	portable analyzer	soil and rock type identification	~0.5% K*	also see under Ca	*63,80*
Rubidium	10 ppm to 5%	solutions	^{125}I, 5 mCi	RbK	...	Si(Li), fwhm 350 eV	laboratory	sensitivity tests	...	matrix effects studied; also see under Br, Mo, and Sr.	*64,65, 66*
Silicon	50 to 70% SiO$_2$	silicates	^{55}Fe, 300 μCi	SiK	...	proportional	laboratory	feasibility study	0.9% SiO$_2$*	matrix and particle size effects studied	*120,132*
	12 to 16% SiO$_2$	sinter mix	^{3}H/Zr, 4 Ci	SiK	...	proportional	on-line analyzer (pressed pellets)	blast furnace control	0.17% SiO$_2$	also see under Fe and Ca	*109,136*
	14% SiO$_2$	cement raw mix	^{3}H/Zr, 3 Ci and ^{210}Po, 10 mCi	SiK	...	proportional	on-line analyzer (fused pellets)	kiln-feed control	0.13% SiO$_2$	also see under Al, Ca, and Fe	*28,129*
Silver	0.01 to 0.1%	silver-bearing ores	^{147}Pm/Al, 0.5 Ci	AgK	Mo/Rh	scintillation	portable analyzer	field assay	0.003% Ag		*112*
	0.01 to 0.1%	silver-bearing ores	^{125}I, 10 mCi	AgK	...	scintillation	laboratory	field assay	0.003% Ag		*19*
	0.01 to 0.1%	silver-bearing ores	^{241}Am-I, 14 mCi	AgK	...	Si(Li), fwhm 580 eV at 22 keV	laboratory	field assay	0.002% Ag	matrix effects studied	*18,112*

Element	Concentration	Sample	Source	Line	Filter	Detector	Instrument	Application	Sensitivity	Remarks	References
	0.01 to 0.5%	artificial ore samples	^{125}I, 10 mCi	AgK	...	Si(Li)	laboratory	field assay	0.01% Ag	feasibility study	18,124
	0.01 to 1%	artificial ore samples	$^{147}Pm/Al$ 2 x 1 Ci	AgK	Mo/Rh	scintillation	borehole logging probe	feasibility study	0.015% Ag	prototype probe	118
Strontium	5 ppm to 5%	solutions	^{125}I, 5 mCi	SrK	...	Si(Li), fwhm 350 eV	laboratory	sensitivity tests	...	matrix effects studied; also see under Br, Mo, and Rb	64,65,66
Sulphur	0.1 to 5% S	pulverized coal	$^{3}H/Ti$, 5Ci	SK	P/S	scintillation	portable analyzer	feasibility study	0.17% S	low noise photomultiplier used	113
	0.01 to 0.5% SO_2	stack gases	^{55}Fe, 10 mCi	SK	P/S	proportional	on-stream analyzer	feasibility study	0.03% SO_2	comparison of nuclear and nonnuclear techniques	116
Tin	0.05 to 25%	tin ore pulps	$^{147}Pm/Al$, 0.5 Ci	SnK	Ag/Pd	scintillation	portable analyzer	mine control	0.03% Sn		12,13, 33,107
	0.5 to 5%	tin mineralization	$^{147}Pm/Al$, 0.5 Ci	SnK	Ag/Pd	scintillation	portable analyzer	in situ assay of rock faces	0.1% Sn		30,33,34
	0.5 to 5%	tin mineralization	$^{35}S/Ba$	SnK	...	scintillation	portable analyzer	in situ assay of rock faces	0.1% Sn		79
	0.5 to 5%	tin ores, cores	$^{147}Pm/Al$, 0.5 Ci	SnK	Ag/Pd	scintillation	portable analyzer	core analysis	0.1% Sn		33,100
	0.002 to 0.1%	tin mineralization	$^{147}Pm/Al$ 0.5 Ci	SnK	Ag/Pd	scintillation	portable analyzer	geochemical prospecting	0.005% Sn	feasibility study	52
	0.05 to 5%	tin mineralization	$^{147}Pm/Al$, 0.5 Ci	SnK	Ag/Pd	scintillation	borehole logging probe	geochemical prospecting; mine control	0.1% Sn		28,51, 92,94
	0.01 to 2%	tin ore slurries	^{241}Am-Ba, ^{241}Am-Sn, 14 mCi	SnK	Ag/Pd	scintillation	on-stream slurry analyzer	process control	0.005% Sn	also see under Mo and Nb	27,48, 107
	0.1 to 11%	bronzes and gun metals	$^{147}Pm/Al$, 0.5 Ci	SnK	Ag/Pd	scintillation	portable analyzer	alloy sorting	0.2% Sn*		108
	0.04 to 8%	copper alloys	^{241}Am, 10 mCi	SnK	...	Si(Li) fwhm 1.2 keV ion chamber	laboratory	feasibility study	0.01% Sn		59
	0 to 1 μm	electroplated tin on steel	$^{3}H/Zr$ 25 Ci	FeK	...	scintillation	on-line coating gage	tin thickness control	0.02 μm*		22
	0 to 1 μm	electroplated tin on steel	^{241}Am-Cs, 27 mCi	SnK	Ag	scintillation	on-line coating gage	tin thickness control	2μg/cm²		134
Titanium	0.04 to 5% 5 to 60% TiO_2	rock, ore, and concentrate pulps	^{55}Fe, 1 mCi	TiK	Sc/Ti or none	scintillation	portable analyzer	field assay and mine control	0.04% Ti		49,50

TABLE 5—*Analyses* (continued).

Element	Concentration Range	Material	Source and Activity	X-ray Excited	Filters	Detector	Instrument	Application	Detection Limit *(or 1σ precision)	Remarks	References
	0.04 to 1% Ti	steels	^{55}Fe, 1.2 mCi	TiK	Sc/Ti or Sc/I	scintillation	portable analyzer	alloy sorting	0.04% Ti	difficult to discriminate between Ti and V with filters	108
Tungsten	0.5 to 20%	steel	^{238}Pu, 20 mCi	WL	Ni/Cu	scintillation	portable analyzer	alloy analysis	0.36% W*		113
	0.5 to 20%	steel	^{147}Pm/Al, 0.5 Ci	WK	...	scintillation	portable analyzer	alloy analysis			113
	1 to 20%	steel	^{57}Co, 1 mCi	WK	...	scintillation	laboratory	feasibility study	0.1% W		44
	0.05 to 10%	aqueous solution	^{57}Co, 1 mCi	WK	...	scintillation	on-stream analyzer	feasibility study	0.08% W		44
Uranium	0.001 to 1% U$_3$O$_8$	ore pulps	^{109}Cd, 1 mCi	UL	Y/Zr	scintillation	portable analyzer	field assay of "non-equilibrium" ores	0.006% U$_3$O$_8$		114
	0.1 to 10 g/l	solution	^{241}Am-Cd, 1 mCi	UL	...	proportional	laboratory analyzer	routine analysis	0.05 g/l		109
	4 to 200 g/l	solution	^{57}Co, 1 mCi	UK	...	scintillation	laboratory analyzer	routine analysis	0.1 g/l		109
	0.002 to 3% U$_3$O$_8$	artificial ore samples	^{57}Co, 2 × 1 mCi	UK	Ta/Pb	scintillation	borehole logging probe	feasibility study	0.1% U$_3$O$_8$	prototype probe	118
Vanadium	0.01 to 1% V$_2$O$_5$	vanadium ores	^{55}Fe, 10 mCi	VK	Ti	scintillation	portable analyzer	mine control	0.014% V$_2$O$_5$		114
	0.05 to 0.7%	steels	^{55}Fe, 1.2 mCi	VK	Sc/Ti	scintillation	portable analyzer	alloy sorting	0.03% V	also see under Ti	108
	0.01 to 3.5%	steels	^{55}Fe, 20 mCi	VK	Sc/Ti	scintillation	portable analyzer	alloy sorting	0.01% V		113
Yttrium	1 μg to 2.5 mg	thin films	^{125}I, 2 mCi	YK	...	Si(Li)	laboratory	feasibility study	1 μg Y	420-min count	69
Zinc	0.1 to 30%	zinc or pulps	^{238}Pu, 30 mCi	ZnK	Ni/Cu	scintillation	portable analyzer	field analysis	0.04% Zn; 0.17% Zn*	interelement and particle size effects studied	13,34, 101
	1 to 20%	zinc ore pulps core samples	^{109}Cd ^{238}Pu, 30 mCi	ZnK ZnK	Ni/Cu	scintillation	... portable analyzer	... core analysis	3% Zn*	...	54 100

0.1 to 30%	lead-zinc and copper-lead-zinc ores, tailings and concentrates	^{238}Pu, 30 mCi	ZnK	Ni/Cu	scintillation	on-stream slurry analyzer	mill control	0.02 to 0.5% Zn*	matrix effects studied	43
50 to 350 g/m²	hot-dipped galvanized steel	^{241}Am, 100 mCi	ZnK	Al	proportional	on-line coating thickness gage	coating control	5% relatives*		89
50 to 350 g/m²	hot-dipped galvanized steel	^{241}Am, 100 mCi	ZnK	Ni/Cu	balanced differential ion chamber	on-line coating thickness gage	coating control	2% relative*		45,119
Zirconium 0.15 to 2%	sands (with 5 to 22% Fe)	^{109}Cd, 0.1 mCi	ZrK	Mo or none	proportional	laboratory	feasibility study	0.01% Zr	matrix and particle size effects studied	60,102
0.1 to 5%	solution	^{147}Pm/Al-Mo and Ru	...	Sr/Zr/Mo	scintillation	laboratory	feasibility study	...	also Hf; absorption edge analysis	23
Miscellaneous:										
Rubber 1.5 mm thick	flooring material	^{241}Am, 1.3 mCi	AgK	...	scintillation	portable gage	wear measurement	±3 μm		68
Iodine 0.04%	thyroid, in vivo	^{241}Am, 3.3 Ci	IK	...	Si(Li) fwhm 650 eV proportional	clinical	flourescent scanning	...		56,57,58
Zinc 1 mg/cm²	postage stamps	^{147}Pm/Al 2 X 2 Ci	ZnK	...	proportional	laboratory on-line gage	automatic letter facing	...		32
Water 300 to 3000 g/m²	paper web	^{241}Am-Ga, 45 mCi	CuK	...	proportional	on-line gage	control of water removal on Fourdrinier machine	2% relative*		73

complex L and K X-ray groups above about 10 keV and $K\alpha_1$, $K\alpha_2$ doublets above about 40 keV. The last achievement exceeds the performance of many diffracting crystal monochromators.

Another aspect of energy resolution is worth mentioning. The higher the resolution of a given peak, the greater is the sensitivity of analysis since in the limit there is always some continuous background under the peak. One advantage of monochromatic sources is that the scattered background can be concentrated far enough away from the signal peak to permit even a low resolution detection system to obtain parts-per-million analytical sensitivity. As the resolution of Si(Li) spectrometers improves, therefore, we look forward to obtaining particularly high sensitivities by using them with spectrally pure line sources.

Balanced Filters [23,109,117]

X-ray filters take the form of thin foils placed over the detector window. By thickness adjustment the X-ray transmissions through two filters of adjacent or close atomic number can be made very nearly equal over a wide range of energies except for a narrow "pass band" between their absorption edges. Figure 10 shows the transmission curves for a nickel/copper balanced filter pair. It is seen that zinc $K\alpha$ X-rays are in the pass band and, therefore, can be separated from possible interfering X-rays, except for copper $K\beta$-lines. Isolation of the X-rays in the pass band is accomplished by measuring the radiation from the specimen transmitted through each filter, in turn, and subtracting one from the other.

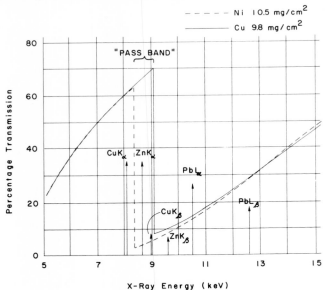

FIG. 10—*X-ray transmission through balanced filters of copper and nickel.*

The measurements can be made with the same detector by changing, oscillating, or rotating the two filters. Alternatively, the outputs from twin measuring heads, each with its respective filter, can be continuously subtracted. This mode is preferred for on-stream analysis.

The criterion for balance at any given energy is that the transmission should be equal at that energy, that is, $\exp - (\mu_a m_a) = exp - (\mu_b m_b)$ or

$$m_a/m_b = \mu_b/\mu_a \dots\dots\dots\dots\dots\dots\dots\dots\dots\dots (1)$$

where m_a, m_b (g/cm^2) are the values of mass per unit area for the filters a and b, and μ_a, μ_b are the corresponding mass attenuation coefficients (cm^2/g).

The optimum mass per unit area of filters is not critical and is given by

$$m_{opt} = \frac{1.5 \pm 0.5}{(\mu_2 - \mu_1)} \ln \frac{\mu_2}{\mu_1} \text{ g/cm}^2 \dots\dots\dots\dots\dots\dots (2)$$

where μ_2 and μ_1 are the mass attenuation coefficients for one of the filters at the top and bottom of its absorption edge.

Filters can be made from metal foils, or the powdered element or oxide can be encapsulated in epoxy resin or plastic and rolled, cast, or pressed into approximately the correct thickness. Where metal foils are used, suitable thicknesses are 0.0005 to 0.001 in. (for titanium to zinc), 0.002 in. (for zirconium to molybdenum), 0.003 in. (for rhodium to tin) and 0.005 to 0.010 in. (for rare earths and hafnium to uranium). Techniques for making plastic or epoxy filters have been described [15,35,113].

A pair of filters can be balanced for a given analysis to a degree of perfection limited mainly by the pains taken. A suitable practical criterion for balance is when the difference count from a "blank" specimen approaches the value of one standard deviation on the same difference count. That is,

$$N_1 - N_2 = \sigma(N_1 - N_2) = (N_1 + N_2)^{1/2} \dots\dots\dots\dots\dots (3)$$

where $\sigma(N_1 - N_2)$ is the standard deviation on the difference between the two counts N_1 and N_2.

Filters are balanced by etching, sanding, or hot pressing (of foil, epoxy, or plastic filters, respectively). For precise balancing, it is often necessary to add a third filter which does not have an absorption edge in the energy region of interest (for example, aluminum foil). Checking the degree of balance is best performed by measuring the transmissions of the X-ray required to be balanced out. One method is to use pure element specimens to check the balance at their characteristic X-ray energies. Two basic problems militate against perfect balancing, unequal absorption-edge ratios and unequal excitation of filter characteristic X-rays by absorbed higher energy radiation. The larger absorption-edge ratio can be effectively reduced by adding the previously mentioned third filter. One filter X-ray intensity can be reduced preferentially by adding the third filter to the detector side of the filter requiring adjustment.

Feasibility Studies and Methods

At this stage in the development of radioisotope X-ray analytical techniques and instruments, only a small proportion of possible analyses have been investigated, so the demand for feasibility studies is relatively high.

One of the main aims of this review is, therefore, to present potential users of the technique with enough information to enable them to decide whether a proposed measurement is feasible, and, if so, what is the simplest apparatus that will perform it.

Basic Equations

The equations for fluorescent and scattered flux derived below have been found to be reasonably accurate and useful in feasibility calculations. The simple model shown in Fig. 11 is used as a basis for their derivation. Two assumptions are made. First, the incident and emitted radiation enter and leave the specimen normally. The broad beam geometries used ensure a wide range of incident and emitted angles. We have found by experience that the normal incidence assumption is satisfactory for central, side, and annular source geometries. The second assumption is that the penetration depth of radiation into the specimen is small compared to the radiation path length from source to specimen and from specimen to detector. This enables the geometrical and radiation interaction factors to be integrated separately, and greatly simplifies the equations. This assumption begins to break down above about 50 keV especially for low density material, but feasibility calculations are still reasonably accurate up to at least 100 keV.

The fluorescent flux from an element a in a layer of mass per unit area $\delta x(g/cm^2)$ at a depth $x(g/cm^2)$ is $\delta I_{Fa} = kI_o\,\omega\,\tau\,r_a\,\delta x\,e^{-\mu_1 x}e^{-\mu_2 x}$ where k is the overall geometrical and detector efficiency [117], r_a is the weight fraction of the element 'a', ω is the fluorescence yield of element a for the X-ray excited (usually ω_K for K X-rays), and τ is the photoelectric absorption coefficient at the incident energy for the element a in the electron shell characterizing the X-ray emitted. The other symbols are described in Fig. 11. The units of all the attenuation cross sections are cm^2/g.

Integrating for a specimen of mass per unit area, m:

$$\frac{I_{Fa}}{kI_o} = \frac{\omega\tau r_a}{(\mu_1 + \mu_2)}\,[1 - \exp - (\mu_1 + \mu_2)m]\dots\dots\dots\dots(4)$$

Note that the values of μ refer to the whole specimen and can be found by a linear summation over all the elements present; for example,

$$\mu_1 = \sum_i \mu_{1i} r_i\dots\dots\dots\dots\dots\dots(5)$$

where r_i is the weight fraction of the ith element. The same applies to μ_2, μ_3, μ_{coh}, and μ_c.

The corresponding equation for the coherent scattered radiation is

$$\frac{I_{coh}}{kI_o} = \frac{\mu_{coh}}{2\mu_1} [1 - \exp - (2\mu_1 m)] \dots\dots\dots\dots\dots(6)$$

Note that the incident and coherent scattered energies are equal.

The Compton backscattered flux is

$$\frac{Ic}{kI_o} = \frac{\mu_c}{(\mu_1 + \mu_3)} [\exp - (\mu_1 + \mu_3)m] \dots\dots\dots\dots(7)$$

The Compton backscattered energy (E') is given by

$$E' = E(1 + 2E/511)^{-1} \text{ keV} \dots\dots\dots\dots\dots(8)$$

where E is the incident energy. Since we assume normal incidence, E' is assumed to be the average Compton scattered energy. This is reasonable at energies below about 50 keV since the total Compton energy change is small.

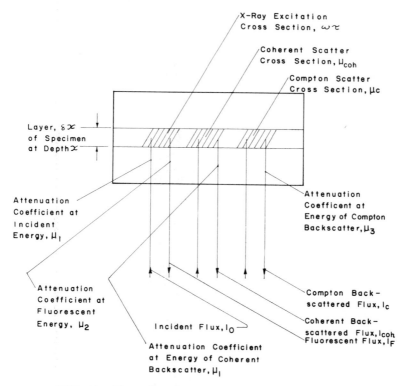

FIG. 11—*Physical model and glossary for basic equations.*

Thick and Thin Specimens—For thick specimens, Eqs 4, 6, and 7 are simplified by equating to unity the portions in square brackets. The criterion chosen for infinite thickness depends on the nearness of the approach to unity desired. A usual criterion is

$$(\mu_1 + \mu_2)m = 4.6 \dots \dots (9)$$

In this case the signal is 99 percent of that from an infinitely thick specimen. Since μ_2 is usually greater than μ_1 and μ_3, a specimen infinitely thick for fluorescent radiation may not be for backscattered radiation. This may cause errors if the backscattered flux is assumed independent of specimen thickness, or may be used to advantage to improve the fluorescent to scatter ratio.

When $(\mu_1 + \mu_2)m \lesssim 0.1$, the specimen is considered to be thin and the exponential functions can be expanded to their first order term. Equations 4, 6, and 7 then become, respectively:

$$I_{Fa}/kI_o = \omega\tau r_a m = \omega\tau m_a \dots \dots (10)$$

where m_a is the mass per unit area of the element a;

$$I_{\text{coh}}/kI_o = \mu_{\text{coh}}m, \dots \dots (11)$$

and

$$I_c/kI_o = \mu_c m. \dots \dots (12)$$

The fluorescent flux is now linearly dependent on the weight concentration of the element. With thick specimens this is not so since the terms μ_1 and μ_2 are dependent upon the concentrations of other elements in the material. This gives rise to the matrix absorption effect in thick specimens, and $(\mu_1 + \mu_2)$ usually is called the matrix absorption term.

Consideration of the above equations leads to the result that although the fluorescent flux from a thin specimen is about an order of magnitude less than that from a thick one, the fluorescent to scatter ratio can be some four times greater (the usual exciting energy is about twice the fluorescent energy, for which $\mu_1 = \mu_3 = \mu_2/7$ at low concentrations of the fluorescent element). Thus when scattering is a problem, higher sensitivity may be possible by analyzing thin specimens.

Polychromatic Exciting Radiation—For sources emitting line spectra with more than one line, the above equations need be calculated for only the one or two most important lines. For bremsstrahlung emitters it is possible to use a single effective energy which is between the absorption edge energy of the element being excited and the maximum bremsstrahlung energy. Calculation of this energy is complicated by the shape of the bremsstrahlung spectrum, which can vary from source to source, and by the nonlinearity of the excitation and absorption functions involved. Alternatively, solution of Eqs 4, 6, and 7 for a number of increments of the bremsstrahlung spectra would give fairly precise results but is rather tedious.

Feasibility Calculations

The above equations can be used for calculating count rate, sensitivity of count rate to concentration, fluorescent to scatter ratio, and matrix absorption effects in homogeneous materials [117]. Matrix enhancement effects are not allowed for, and suitable equations have not been derived, to our knowledge, for radioisotope X-ray spectrometry. However, the effects are second order, and one need not be too concerned with them in a feasibility calculation.

Lubecki [84] has calculated the effect of the source energy on the sensitivity of analysis and concludes that other things being equal, the sensitivity should increase with increasing exciting energy. However, instrumental factors such as detector resolution and polychromatic sources can reverse this conclusion. Leman [77,78] also discusses the choice of optimum excitation energy.

Lubecki [85] has derived a general form of sensitivity function suitable for calculating the sensitivity of multiple, interdependent measurements.

Matrix Effects

Interferences due to internal absorption and enhancement of the fluorescent radiation and to changes in emitted intensity due to specimen heterogeneity and finite particle size are functions of the specimen and so are basically the same whatever X-ray fluorescence analysis technique is used. Minimization of errors due to these effects is a question of central importance in both conventional and radioisotope X-ray spectrometry.

The methods used to minimize matrix effects in conventional X-ray spectrometry also can be used with the radioisotope technique as long as they do not demand resolutions or X-ray fluxes higher than are possible with the radioisotope technique.

A number of methods for minimizing matrix absorption effects have been tested for use with portable and on-stream analyzers. Many of these were evaluated in the early 1960's, and the following ones were reviewed in 1966 [109]: (1) the use of balanced filters to isolate the fluorescent radiation from scattered radiation and so limit the effect to that of the fluorescent radiation alone; (2) compensation for the decrease in intensity of the characteristic X-rays from the wanted element by exciting and detecting, simultaneously, X-rays from the interfering element; (3) the use of fluorescent to backscattered ratio; (4) specimen dilution in "on-stream" analysis; (5) the use of fluorescent to transmission ratio (especially in on-stream analysis); (6) the use of two energies, one that excites the required characteristic radiation and one that does not [11,25], and (7) the use of empirically determined nomograms to solve two-component problems.

Since 1966, the use of fluorescent to backscattered ratio [81,103,104, 112] and of nomographic [99,125] techniques has increased, particularly for applications of portable analyzers [28]. The emission-transmission method has been evaluated for use with radioisotope sources by Lubecki [88]. In this method the specimen is backed by a target, and the absorption of the characteristic X-rays of the latter in the specimen is used to measure the matrix effect. Dziunikowski [41] has proposed a method using four balanced filters and a continuous source spectrum to eliminate two kinds of interference in the analysis of copper ores, one due to variation of iron content and the other to changes of rock type. Perhaps the most comprehensive recent paper is a theoretical review and comparison by Lubecki [87] of six methods of eliminating or reducing matrix effects in radioisotope X-ray fluorescence and absorption analysis. The methods reviewed are dilution, fluorescence-absorption, fluorescence-backscatter, emission-transmission, double channel absorption edge analysis, and multicomponent analysis.

Particle Size Effects—Two new theories have been developed to explain the effects of finite particle size on the fluorescent intensity excited in a particulate specimen [7,82]. These theories represent a significant advance on older explanations of particle size effects. For example, they can account for certain simple phenomena such as the variation of fluorescent intensity with packing density, and with particle size in one component systems, whereas older theories could not. The Berry-Furuta-Rhodes theory [7] appears to make more realistic assumptions about random packing of grains and has predicted correctly some surprising changes in fluorescent intensity with particle size in complex systems. It also has been extended to explain the particle size dependence of the backscattered and transmitted X-ray intensities. All equations describing the effects of particle size are complex and cumbersome to enumerate. Lubecki has derived a nomographic method for fast estimation of the particle size effect which should prove useful in feasibility calculations [83].

The new theories ought to be as effective in predicting particle size effects in conventional X-ray spectrometry as they have been to date in radioisotope X-ray spectrometry.

Applications and Instruments

Significant analyses reported in the literature are classified alphabetically by element in Table 5. The main objective of this classification is to provide easy reference not only to what analyses have been studied but also to the apparatus in use and the conditions of applications. Forty-five percent of the 116 analyses tabulated use standard portable analyzers, in environments divided fairly evenly between field, industrial, and laboratory conditions. Another 5 percent of the analyses use borehole logging probes. Laboratory feasibility studies or routine laboratory determinations constitute 25 percent of the applications, of which half use Si(Li) or Ge(Li) detectors. The rest of the applications, 24 percent, are to industrial process control. These include "on-stream" analyzers, where a continuously moving sample of the plant stream is measured, and instruments which measure discrete samples taken automatically.

It is not known how many individual instruments are in use, but it is probable that the number of portable analyzers in use is much greater than the number of papers written on their use, since they have been commercially available for five years. However, other analyzers are not yet generally available, and the number in use is not likely to greatly exceed the number of papers written.

Study of the research investigations indicates that in about 60 percent of them the objective was to optimize the accuracy of major component determinations, and, in the rest, it was to obtain the best possible sensitivity. In the former case, the main sources of error are matrix effects. In applications of portable analyzers, much ingenuity has gone into devising methods of minimizing these errors without compromising the essential simplicity of the equipment and measurement procedures. We expect to see more development along these lines, as matrix effects appear to be one of the main obstacles to much more widespread field use.

Sensitivity is limited by the basic considerations of counting statistics and fluorescent-to-scatter ratio. Using scintillation and proportional counters, detection limits of about 0.005 percent (and in favorable cases, 0.001 percent [106]) can be usually obtained by one of two methods. The first is to use a source whose backscattered radiation is just resolved from the required characteristic X-rays. The second is to excite the required fluorescence with maximum possible efficiency using an energy just above the absorption edge of the element to be determined. Resolution limitations prevent employment of both these methods simultaneously with proportional or scintillation counters. They do not, however, when Si(Li) or Ge(Li) detectors are used, and in this case sensitivities of a few parts per million can be obtained even with heterogeneous ore specimens [18,112]. It seems that the development potential of proportional and scintillation counters in this respect has been thoroughly investigated and largely exhausted [19]. This is not the case with Si(Li) and Ge(Li) spectrometers where many studies using obvious combinations of high resolution detectors, monochromatic line sources, and thin specimens (for example) have not yet been performed. We expect to see a rapid increase in the number of investigations in this area.

Portable Analyzers

Portable analyzers are the first and only class of developed radioisotope X-ray fluorescence instruments that have been in routine use long enough for systematic evaluations of a range of analyses to be available [13,28,34,

49,51]. Applications have been mainly to pulp, core, and rock face analyses in mining, but other significant applications are to alloy and process solution analyses. Elements from calcium to uranium can be determined with standard equipment.

The first such instruments were made available by British manufacturers in 1965 to 1967, and most of the reported applications have been with these instruments. Some half dozen instruments are now available in the United States, and reports of experiences with them should appear in the literature over the next few years. A number of papers describing these instruments and feasibility studies made with them have been published already [*36,63,80,110,111,113-115,128*]. Also, in the last two years a good deal of Russian work has been published [*4,54,79,97,98,104,125*].

The first instrument [*12,33,107,108*] consisted of a 2-in.-diameter NaI (T1) scintillation probe with balanced filters, and an electronic unit which included an amplifier, single discriminator, and ratemeter readout. Most of the instruments developed since then have used similar scintillation probes, but the sophistication and performance of the electronic units have been improved. They include a single channel analyzer with variable window width and an up-down scaler for convenient display of the difference count from the balanced filters. Count times cover the range 10 to 100 s, which is the usual range of convenience for the type of measurement undertaken. A great deal of flexibility in power supplies is available including line operation, and operation from automobile batteries, rechargeable batteries, and dry cells. Battery life varies from 10 to 50 h. A photograph of one of the latest instruments is shown in Fig. 12. This has a smaller scintillation probe than usual, 1.5 in. diameter, and employs side source geometry. The manufacturer claims that improvements in the geometrical design have enabled significantly better sensitivities to be obtained using smaller specimens than hitherto.

Although proportional counters have better energy resolution than scintillation counters, their application to portable analyzers has been restricted to one or two instruments [*4,36,104,128*]. The probable reason for this is the high sensitivity of pulse height to count rate at the high count rates (10^3 to 10^5/s) used in these instruments. Recently attempts have been made to solve this problem by using a reference source and gain stabilizer [*4,36*].

Avalanche detectors [*63*] also have been developed to the point where they could be used in portable analyzers. They ought to find applications where their ultra-small size outweighs their poor energy resolution. High resolution Si(Li) and Ge(Li) detectors have not yet found significant field application, though Frierman [*47*] reports their use on an archeological expedition.

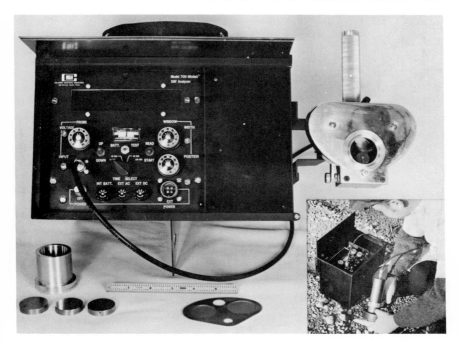

(Courtesy of Columbia Scientific Industries Corporation)

FIG. 12—*Photograph of portable radioistope X-ray fluorescence analyzer*

Borehole Logging Probes

Borehole logging for elemental analysis has not found wide use in mining because of the lack of suitable analysis techniques. Coring followed by core analysis is widespread but can be an order of magnitude more expensive than plain drilling, especially now that new rock drilling methods are available. Although restricted to dry, uncased holes, and to determination of relatively high atomic number elements, radioisotope X-ray fluorescence drill hole probes have been developed and are beginning to find applications in mine workings for delineation of ore bodies and measurement of ore grade [2,28,118]. Figure 13 shows a typical probe schematic. Two detectors are preferred if balanced filters are to be used, so as to avoid moving parts in the probe.

Measurements have been restricted largely to elements of atomic number greater than about 45 because of the relatively low penetration of X-rays below 20 keV through the necessarily rugged probe windows. Most of the work reported has been performed in Russia. Kudryavtsev [71], Meyer [93,95] and Mitov [96] describe research studies. Alekseyev [2], Meyer [92], and Leman [75,76] describe the assay of antimony ores, and Meyer

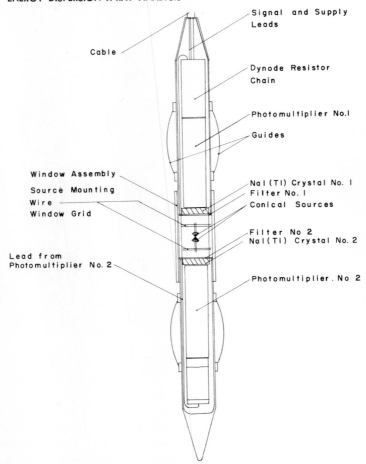

FIG. 13—*Possible schematic of X-ray fluorescence borehole logging probe.*

[*94,95*] also describes logging for tin content of tin ores. Clayton [*28*] reports tests of a probe with "Melinex" or "Duralumin" windows for assay of copper and tin ores, respectively. Work reported in the United States describes feasibility studies on silver and uranium determination using a prototype probe and artificial boreholes [*118*].

Laboratory Analyses

Laboratory analyses have been confined largely to feasibility studies and special applications. No standard design of instrument is available, although Robert [*120*] has recently described an instrument for light element analysis. Reference should be made to the entries in Table 5 for detailed information on applications and instrumentation. Nearly all the reports on the properties and uses of Si(Li) and Ge(Li) detectors fall into this category. When they were first proposed for use in X-ray fluorescence analysis [14,61,69], the resolution of Si(Li) and Ge(Li) detectors in the important energy range 5 to 10 keV was somewhat worse than that of proportional counters, but their potential was envisaged clearly. Since then a series of papers has been published [1,17,55] describing the rapid advances made in detector and electronic technology, which has brought us to the point illustrated in Fig. 9, where these detectors can now resolve multicomponent spectra from adjacent elements. One point of practical importance which should find favor with instrument designers is the extreme stability of Si(Li) and Ge(Li) spectrometers. Energy-pulse height calibrations can be maintained to a fraction of one percent over periods of several months without special precautions. The reasons for this are as follows: the detector and preamplifier are temperature-controlled by the liquid nitrogen; the detector output pulse height is nearly independent of bias voltage; and circuitry after the linear pulse amplifier is digital.

On-Stream Analyzers

Clayton [28] recently has reviewed developments of on-stream radioisotope X-ray analyzers in the mineral processing industry. The most progress appears to have been made in Australia and Britain. Australian equipment uses scintillation counters and source-target assemblies, with applications mainly to determination of copper, lead, zinc, and tin in flowing slurries, for control of mineral processing plants [42,43,135]. Similar instruments have been developed and installed in Britain on mineral processing pilot plants for determination of zinc, niobium, molybdenum, tin, and barium [27]. Special flow cells to minimize variations of X-ray intensity with slurry density also have been developed [24], and on-stream radioisotope X-ray techniques for correction of particle size effects have been demonstrated [26]. White [136] describes an instrument for on-stream determination of silicon, calcium, and iron in automatically sampled and pelletized iron ore sinter mix. On-stream analysis of cement raw mix has received a good deal of attention [10,70,86,127,129]. Boirat [10] describes the determination of calcium and iron in a moving sample stream of powdered cement raw mix. Uchida [129] describes an instrument that uses combined alpha particle and X-ray excitation to determine aluminum, silicon, calcium, and iron in automatically-sampled fused pellets of cement raw mix. Sampling and analysis is claimed to take only 10 min. Starnes [127] compares continuous presentation of dry powders and aqueous slurries and describes an instrument for on-stream determination of calcium in cement raw mix slurries.

On-line radioisotope X-ray fluorescence analyzers also have been developed for continuous measurement of the thickness of electroplated and hot-dipped tin and zinc coatings on steel strip. These are true "on-line"

analyzers since the measurement is made in "real time" on the actual coating line. The first such application was to electroplated tin and zinc on
steel [22]. This instrument uses multiple ^{3}H/Zr sources to excite iron K
X-rays in the steel. The decrease in their intensity with increasing tin or
zinc thickness is monitored by a thin-windowed industrial ionization
chamber. Watt [133] describes the use of source-target assemblies for determination of tin coating weight on steel by excitation of tin K X-rays.

Excitation of iron K X-rays cannot be used for measurement of hot-
dipped zinc coatings on steel strip because they are attenuated almost to
background levels by the relatively thick coating. Instead, zinc K X-rays
are excited with 60 keV ^{241}Am gamma rays [42,89,119]. A double ion
chamber having two adjacent windows each covered with one of a balanced filter pair is used to isolate the zinc $K\alpha$ X-radiation. The two halves
of the ionization chamber are polarized in opposite senses; hence, the
output current is the required difference between the filter transmissions
[42,119].

Kuusi [73] describes a novel method for monitoring the water content at
the wet end of the paper web on a paper-making machine. Gallium K
X-rays from an ^{241}Am-Ga source-target assembly shine through the web
and excite copper K X-rays in the support grid. These are detected by a
proportional counter after emerging from the web on the same side of the
source. At this point in the paper-making process, the quantity of fibers
and fillers is much less than that of water and is effectively constant. Also,
the mass attenuation coefficient of water at 8 keV is some 50 percent
greater than that of paper fiber.

Perhaps the fastest radioisotope on-stream X-ray analyzer is that developed by Darigny [32] for detecting the presence of a stamp on an
envelope in an automatic letter-orientation machine. The stamp adhesive
contains a few percent of zinc whose K X-rays are excited by a ^{147}Pm/A1
source and detected by a proportional counter. The instrument can sort
36,000 letters per hour with less than one mistake in 10,000. The equipment is being considered for installation by the French Post Office instead
of conventional ultraviolet techniques.

New Applications

Four applications deserve special mention. They are the uses of radioisotope X-ray fluorescence analysis in the measurement of wear [68], in
criminalistics [122], in medical scanning [56-58] and as an educational
tool [121]. Three of these involve the use of an inactive tracer containing
an element whose characteristic X-rays are excited for the measurement.
The greatest advantage of this, compared with the use of radioactive
tracers, is that radiation is used only during the actual measurement.

Kemper [68] describes a method for monitoring the wear rates of flooring materials by incorporating a target sheet just behind the wearing layer. The energy of the source and the characteristic energy of the target must be chosen so that the degree of attenuation of the incident and emitted radiation is suitable for measuring the required thickness range. For rubber flooring materials 1.5 mm thick, the source chosen was ^{241}Am and the target, silver. With the apparatus developed, thickness could be measured to within 3 to 20 μm in a 5-min count time. Standard portable analyzers can be used for this application. Many different applications to wear measurement can be envisaged, particularly when access is possible to one side only of the material.

Sellers [122] describes possible uses of inactive tracers in criminalistics for covert marking of a wide range of materials for authentication purposes, and for bullet hole identification. The advantage of the X-ray method in the first application is the difficulty of discovering and elucidating the marking code. In both applications, portable analyzers could be used as the readout probe. A prototype portable bullet hole identifier is being tested.

Scanning techniques for imaging internal organs of the body are now standard practice in medical diagnostics. The most widely used method is to ingest a radioactive tracer whose biochemical function results in localization at the proper site and whose emission of gamma rays enables external, *in vivo* visualization of that site using gamma cameras or scanners. In "fluorescent scanning" [56-58] the ingested chemical is not radioactive but contains an element whose characteristic X-rays can be excited and detected by an external probe. Thus, the patient only receives a radiation dose during the measurement instead of the whole time that the radiotracer is in the body. Satisfactory thyroid scans have been obtained by measuring iodine K X-rays using ^{241}Am and a Si(Li) detector. An order of magnitude reduction in whole body dose is obtained compared with the radiotracer method.

The availability of inexpensive radioisotope X-ray equipment makes it possible for educational establishments, who hitherto could not afford it, to provide courses, demonstrations, and experiments in atomic and nuclear physics, and on the interaction of X-rays and gamma radiation with matter. Robinson [121] presents an excellent description of a range of experiments in basic radioactivity, X-ray physics, spectrometery, thickness measurement, and analysis that can be performed with a proportional, scintillation, or solid-state detector, a few radioisotope sources and some element foils to act as targets, specimens, or filters. To our knowledge the potential of this type of apparatus in education, especially at the junior college and undergraduate level, has not been recognized at all.

References

[1] Aitken, D. W., "X-ray Detection in Semiconductors and Scintillators," Final Report AFOSR 69-1867 TR, Jan. 1969.

[2] Alekseyev, V. V., Ochkur, A. P., and Suvorow, A. D., "Application of X-ray Fluorescence and Photoneutron Methods in Studying Elemental Composition of Rocks and Ores," *23rd International Geology Congress*, Academia, Prague, 1968, pp. 57-65.

[3] Ansell, K. H. and Stevenson, J., in ORNL-IIC-10, Baker, P. S. and Gerrard, M., eds., Oak Ridge National Laboratory, Oak Ridge, Tenn., Sept. 1967, pp. 764-785.

[4] Baronin, V. N. et al, *Zavodskaia Laboratoriia*, Vol. 30, 1964, No. 4, pp. 498-500.

[5] Bednarek, B., Jelen, K., and Ostrowski, K., *Nukleonika*, Vol. 12 (1-2), 1967, pp. 135-141.

[6] Berry, P. F., *Transactions of the Canadian Institution of Mining and Metallurgy*, Vol. LXX, 1967, pp. 98-106.

[7] Berry, P. F., Furuta, T., and Rhodes, J. R., in *Advances in X-ray Analysis*, Vol. 12, Newkirk, J. B., Mallett, G. R., and Pfeiffer, H. G., eds., Plenum Press, New York, 1969, pp. 612-632.

[8] Birks, L. S., *Applied Spectroscopy*, Vol. 23, 1969, pp. 303-348.

[9] Bochenin, V. I., *Zavodskaia Laboratoriia*, Vol. 33, 1967, pp. 1158-1159.

[10] Boirat, Von R., "Anwendung von Radioelementen zur Kontinuierlichen Mengenbestimmung Chemischer Elemente," *Dechema Monographien Band 61*, Nr 1083-1101, Verlag Chemie, Gmbh, Weinheim/Bergstrasse, 1968.

[11] Bol'shakov, A. Yu., *Atomnaia Energiia* (USSR) Vol. 25, No. 6, 1968, pp. 535-536.

[12] Bowie, S. H. U., Darnley, A. G., and Rhodes, J. R., *Transactions of the Institution of Mining and Metallurgy*, Vol. 74, 1964-1965, pp. 361-379.

[13] Bowie, S. H. U., *Mining Magazine*, London, Vol. 118, 1968, pp. 1-6.

[14] Bowman, H. R., Hyde, E. K., Thompson, S. G., and Jared, R. C., *Science*, Vol. 151, Feb. 1966, pp. 562-568.

[15] Broquet, C., Robin, G., and Vacher, M., in ORNL-IIC-10, Baker, P. S. and Gerrard, M., eds., Oak Ridge National Laboratory, Oak Ridge, Tenn., Sept. 1967, pp. 356-375.

[16] Brunner, G., *Isotopenpraxis*, Vol. 3, 1967, pp. 165-174.

[17] Burkhalter, P. G. and Campbell, W. G., in ORNL-IIC-10, Baker, P. S. and Gerrard, M., eds., Oak Ridge National Laboratory, Oak Ridge, Tenn., Sept. 1967, pp. 393-423.

[18] Burkhalter, P. G., in *Nuclear Techniques and Mineral Resources*, IAEA, Vienna, 1969 (STI/PUB/198), pp. 365-379.

[19] Burkhalter, P. G., *International Journal of Applied Radiation and Isotopes*, Vol. 20, 1969, pp. 353-362.

[20] Campbell, W. J. in *X-ray and Electron Methods of Analysis*, van Olphen, H. and Parrish, W., eds., Plenum Press, New York, 1968, Chapter II.

[21] Campbell, W. J. and Brown, J. D., *Analytical Chemistry*, Vol. 40, April 1968, pp. 346R-375R.

[22] Cameron, J. F. and Rhodes, J. R., *British Journal of Applied Physics*, Vol. 11, 1960, pp. 49-52.

[23] Cameron, J. F. and Rhodes, J. R., in *Encyclopedia of X-rays and Gamma Rays*, Clark, G. L., ed., Reinhold, New York, 1963, pp. 150-156.

[24] Carr-Brion, K. G. and Jenkinson, D. A., *Journal of Scientific Instruments*, Vol. 42, 1965, p. 817.

[25] Carr-Brion, K. G. and Jenkinson, D. A., *British Journal of Applied Physics*, Vol. 17, 1966, pp. 1103-1104.

[26] Carr-Brion, K. G. and Mitchell, P. J., *Journal of Scientific Instruments*, Vol. 44, 1967, pp. 611-614.

[27] Carr-Brion, K. G., *Transactions of the Institution of Mining and Metallurgy*, Vol. 76, 1967, pp. C94-C100.

[28] Clayton, C. G. in *Nuclear Techniques and Mineral Resources*, IAEA, Vienna, 1969 (STI/PUB/198), pp. 293-324.

[29] Cox, J. A., *Transactions of the Institution of Mining and Metallurgy*, Vol. 76, 1967, pp. A149-A159.

[30] Cox, R., *Transactions of the Institution of Mining and Metallurgy*, Vol. 77, 1968, pp. B109-B116.

[31] Cranston, F. P. and Anspaugh, L. R., "Preliminary Studies in Nondispersive X-ray Fluorescent Analysis of Biological Materials," UCRL-50569, Lawrence Radiation Laboratory, Livermore, Calif., 1969.

[32] Darigny, E. and Robin, G., in ORNL-IIC-10, Baker, P. S. and Gerrard, M., eds., Oak Ridge National Laboratory, Oak Ridge, Tenn., Sept. 1967, pp. 786-804.

[33] Darnley, A. G. and Leamy, C. C., in *Radioisotope Instruments in Industry and Geophysics*, Vol. I, IAEA, Vienna, 1966, pp. 191-211.

[34] Darnley, A. G. and Gallagher, M. J., *Transactions of the Institution of Mining and Metallurgy*, Vol. 75, 1966, pp. B105-B106.

[35] Dunne, J. A. and Nickle, N. L., in ORNL-IIC-10, Baker, P. S. and Gerrard, M., eds., Oak Ridge National Laboratory, Oak Ridge, Tenn., Sept. 1967, pp. 336-355.

[36] Duftschmid, K. E., in *Nuclear Techniques and Mineral Resources*, IAEA, Vienna, 1969 (STI/PUB/198), pp. 325-342.

[37] Dziunikowski, B., *Transactions of the Institution of Mining and Metallurgy*, Vol. 76, 1967, pp. B202-B209.

[38] Dziunikowski, B. and Skrzeszewski, Z., *Nukleonika*, Vol. 12(1-2), 1967, pp. 81-89.

[39] Dziunikowski, B., *Nukleonika*, Vol. 13 (4-5), 1968, pp. 361-369.

[40] Dziunikowski, B. and Niewodniczanski, J. W., in *Nuclear Techniques and Mineral Resources*, IAEA, Vienna 1969 (STI/PUB/198), pp. 343-351.

[41] Dziunikowski, B. and Clayton, C. G., Research Group Report AERE-R5914, United Kingdom Atomic Energy Authority, Her Majesty's Stationery Office, London, 1969.

[42] Ellis, W. K., Fookes, R. A., Watt, J. S., Hardy, E. L., and Stewart, C. C., *International Journal of Applied Radiation and Isotopes*, Vol. 18, 1967, pp. 473-478.

[43] Ellis, W. K., Fookes, R. A., Gravitis, V. L., and Watt, J. S., *International Journal of Applied Radiation and Isotopes*, Vol. 20, 1969, pp. 691-701.

[44] Enomoto, S., "Heavy Element Concentration Determination by the X-ray Fluorescence Analysis Using Radioisotope Gamma-Ray Sources," Final Report, CEA-R-3369, CEN, Saclay, 1968.

[45] Floeck, J. H., *Products Finishing*, Vol. 32, No. 9, 1968, pp. 66-70.

[46] Florkowski, T., *Atomwirtschaft, Atomtechnik*, Vol. 12, 1967, pp. 247-254.

[47] Frierman, J. D., Bowman, H. R., Perlman, I., and York, C. M., *Science*, Vol. 164, March 1969, p. 588.

[48] Furuta, T. and Rhodes, J. R., *International Journal of Applied Radiation and Isotopes*, Vol. 19, 1968, pp. 483-485.

[49] Gallagher, M. J., *Transactions of the Institution of Mining and Metallurgy*, Vol. 76, 1967, pp. B155-B164.

[50] Gallagher, M. J., *Transactions of the Institution of Mining and Metallurgy*, Vol. 77, 1968, pp. B129-B134.

[51] Gallagher, M. J., "Portable X-ray Spectrometers for Rapid Ore Analysis," *Ninth Commonwealth Mining and Metallurgical Congress*, May 1969, Institution of Mining and Metallurgy, London.

[52] Garson, M. S. and Bateson, J. H., *Transactions of the Institution of Mining and Metallurgy*, Vol. 76, 1967, p. B165.

[53] Giauque, R. D., *Analytical Chemistry*, Vol. 40, 1968, pp. 2075-2077.

[54] Grinshtein, Yu. A., Burmenckiy, A. P., Romanenko, G. V., Dvorechenskiy, F. I., and Halgeer, O. D., "On the Possibility of Determination of Lead and Zinc Content in Polymetallic Ores by X-ray Radiometric Method," *Problems of Exploration Geophysics*, Vol. 11, Leningrad, 1969, pp. 37-40.

[55] Heath, R. L., in ORNL-IIC-10, Baker, P. S. and Gerrard, M., eds., Oak Ridge National Laboratory, Oak Ridge, Tenn. Sept. 1967, pp. 424-441.

[56] Hoffer, P. B., Jones, W. B., Crawford, R. B., Beck, R., and Gottschalk, A., *Radiology,* Vol. 90, 1968, pp. 343-344.

[57] Hoffer, P. B., Charleston, D. B., Beck, R. N., and Gottschalk, A., in *Medical Radioisotope Scintigraphy,* Vol. I, IAEA, Vienna, 1969, pp. 261-271.

[58] Hoffer, P. B., *Isotopes and Radiation Technology,* Vol. 6, Spring 1969, pp. 292-294.

[59] Hollstein, M. G. and DeVoe, J. R., in ORNL-IIC-10, Baker, P. S. and Gerrard, M., eds. Oak Ridge National Laboratory, Oak Ridge, Tenn., Sept. 1967, pp. 483-502.

[60] Holynska, B. and Langer, L., *Analytica Chimica Acta,* Vol. 40, 1968, pp. 115-122.

[61] Hyde, E. K., Bowman, H. R., and Sisson, D. H., "Analytical Capabilities of a Semiconductor X-ray Emission Spectrometer," UCRL-16845, Lawrence Radiation Laboratory, Berkeley, Calif., 1966.

[62] Jelen, K., Ostrowski, K., and Lasa, J., *Nukleonika,* Vol. 12(1-2), 1967, pp. 77-83.

[63] Johnson, P. A., Huth, G. C., and Locker, R. J., *Isotopes and Radiation Technology,* Vol. 7, 1970, pp. 266-277.

[64] Jones, W. B. and Carpenter, R. A. in ORNL-IIC-10, Baker, P. S. and Gerrard, M., eds., Oak Ridge National Laboratory, Oak Ridge, Tenn., Sept. 1967, pp. 465-482.

[65] Jones, W. B. and Carpenter, R. A., in *Advances in X-ray Analysis,* Vol. 11, Newkirk, J. B., Mallett, G. R., and Pfeiffer, H. G., eds., Plenum Press, New York, 1968, pp. 214-229.

[66] Jones, W. B. and Carpenter, R. A., in *Nucleonics in Aerospace,* Polishuk, P., ed., Plenum Press, New York, 1968, pp. 323-333.

[67] Kato, M., in ORNL-IIC-10, Baker, P. S., Gerrard, M., eds., Oak Ridge National Laboratory, Oak Ridge, Tenn., Sept. 1967, pp. 723-745.

[68] Kemper, A., *Wear,* Vol. 12, July 1968, pp. 55-68.

[69] Klecka, J. F., "Analysis for Yttrium with a Semiconductor X-ray Emission Spectrograph and an I-125 Source," UCRL-17144, Lawrence Radiation Laboratory, Berkeley, Calif., 1966.

[70] Kobyakov, B. S., *Pribory i Sistemy Avtomatiki,* Vol. 6, 1967, pp. 43-45.

[71] Kudryavtsev, Yu. I. and Meyer, V. A., "Influence of Ore Structure on the Intensity of X-ray Fluorescence and Scattered Radiation and their Ratio, in the In-Situ Measurements," *Scientific Papers of Leningrad State University, Ucheny Zapiski L.G.U.,* Vol. 340, Leningrad, 1968, pp. 150-177 (in Russian).

[72] Kuusi, J., *Kemian Teollisuus,* Vol. 25, 1968, pp. 478-484.

[73] Kuusi, J., Hietala, M., Puolakka, H., and Lehtinen, A. I., *Tappi,* Vol. 52, No. 12, Dec. 1969, pp. 2378-2381.

[74] Langheinrich, A. P. and Forster, J. W. in *Advances in X-ray Analysis,* Vol. 11, Newkirk, J. B., Mallett, G. R., and Pfeiffer, H. G., eds., Plenum Press, New York, 1968, pp. 275-286.

[75] Leman, E. P., Ochkur, A. P., and Orlov, V. N., "X-ray Fluorescence Logging of Subsurface Boreholes in Antimony Deposits," *Problems of Prospecting Geophysics, Voprosy Razvedochnoy Geofiziki,* No. 7, Nedra Press, Leningrad, 1968, pp. 42-49 and 156-159 (in Russian).

[76] Leman, E. P. and Kommissarzhevskaya, G. F., "X-ray Fluorescence Sampling of Mine Excavations in Antimony Deposits," *Razvedka i Okhrana Nedr,* No. 1, 1968, pp. 45-48 (in Russian).

[77] Leman, E. P. and Bolotova, N. G., "Radioactive Isotopes and Gamma-Ray Sources for X-ray Sampling of Ores," *Problems of Exploration Geophysics, Voprosy Razvedochnoy Geofiziki,* Vol. 11, Leningrad 1969, pp. 3-11 (in Russian).

[78] Leman, E. P., Ochkur, A. P., Bolotova, H. G., Yanchevskiy Yu. P., and Mimor V. H., "Some Peculiarities and Conditions of Application of Radioisotope Gamma-Sources for X-ray Radiometric Ore Sampling," *Problems of Exploration Geophysics, Voprosy Razvedochnoy Geofiziki,* Vol. 11, Leningrad, 1969, pp. 12-18 (in Russian).

[79] Leman, E. P. and Bolotova, N. G., "Determination of Tin and Lead in Ores in Mine Walls by X-ray Radiometric Method," *Problems of Exploration Geophysics*, Vol. 11, Leningrad, 1969, pp. 41-46 (in Russian).

[80] Locker, R. J. and Huth, G. C., *Applied Physics Letters*, Vol. 9, 1966, pp. 227-230.

[81] Lubecki, A., Wasilewska, M., and Gorski, L., *Spectrochimica Acta*, Vol. 23A, 1967, pp. 831-840.

[82] Lubecki, A., Holynska, B., and Wasilewska, M., *Spectrochimica Acta*, Vol. 23B, 1968, pp. 465-479.

[83] Lubecki, A., *Spectrochimica Acta*, Vol. 23B, 1968, pp. 497-502.

[84] Lubecki, A. and Wasilewska, M., *Journal of Radioanalytical Chemistry*, Vol. 1, 1968, pp. 211-218.

[85] Lubecki, A., *Journal of Radioanalytical Chemistry*, Vol. 1, 1968, pp. 413-417.

[86] Lubecki, A. and Wasilewska, M., *Journal of Radioanalytical Chemistry*, Vol. 1, 1968, pp. 25-27.

[87] Lubecki, A., *Journal of Radioanalytical Chemistry*, Vol. 2, 1969, pp. 3-18.

[88] Lubecki, A., *Journal of Radioanalytical Chemistry*, Vol. 3, 1969, pp. 317-328.

[89] Margolinas, S. in ORNL-IIC-10, Baker, P. S. and Gerrard, M., eds., Oak Ridge National Laboratory, Oak Ridge, Tenn., Sept. 1967, pp. 805-816.

[90] Martinelli, P., *Chimia*, Vol. 21, 1967, pp. 151-159.

[91] Martinelli, P., Robert, A., and Cavailles, J., in ORNL-IIC-10, Baker, P. S., and Gerrard, M., eds., Oak Ridge National Laboratory, Oak Ridge, Tenn., Sept. 1967, pp. 696-722.

[92] Meyer, V. A. and Nakhabtsev, V. S., "Application of X-ray Fluorescence Logging for Quantitative Determination of Heavy Elements in Ores of Complex Composition," *Bulletin of Science-Technical Information*, Ministry of Geology USSR, Series: Regional, Prospecting and Logging Geophysics, Vol. 7, 1967, pp. 13-20 (in Russian).

[93] Meyer, V. A. and Nakhabtsev, V. S., "On the Influence of Multiscattering in X-ray Fluorescence Logging," *Vestnik Leningradshogo Universitela, Seriya Geologii Geografii*, Vol. 22, 1967, pp. 83-88 (in Russian).

[94] Meyer, V. A. and Nakhabtsev, V. S., "Results of Testing of X-ray Fluorescence Logging in Tin Ore Deposits," *Vestnik Leningradshogo Universitala Seriya Geologii Geografii*, No. 18, 1968, pp. 51-56 (in Russian).

[95] Meyer, V. A., Rozuvanor, A. P., and Nakhabtsev, V. S., "Influence of Background Compensation in X-ray Fluorescence Log," *Scientific Papers of the Leningrad State University No. 346; Physical and Geological Series No. 19: Problems of Geophysics (Voprosy Geofiziki)*, Leningrad, 1969 (in Russian).

[96] Mitov, V. N., Yanshevskiy, Yu. P., and Leman, E. P., "Some Singularities in the Secondary Gamma Spectra in the Geometry of Direct Visibility, and Application in the X-ray Fluorescence Method," *Transactions of the Leningrad Mining Institute (Zapiski Leningradshogo Gornogo Instituta)*, Vol. 56, No. 2, 1969, pp. 122-127 (in Russian).

[97] Mitov V. N., "X-ray Radiometric Analysis of Lead in Samples Using a Selenium-75 Source," *Problems of Exploration Geophysics*, Vol. 11, Leningrad, 1969, pp. 26-32 (in Russian).

[98] Mitov V. N., Volfshtein P. M., and Yanshevskiy Yu. P. "Sampling of Drilling Core by X-ray Radiometric Method," *Problems of Exploration Geophysics*, Vol. 11, Leningrad, 1969, pp. 33-36 (in Russian).

[99] Niewodniczanski, J. W., *Nukleonika*, Tom XI, 1966, pp. 725-734.

[100] Niewodniczanski, J. W., "Application of Radioisotope X-ray Fluorescence Analysis for Drill-Core Scanning," *Institute of Geological Science Radiogeology and Rare Mineral Unit*, London, Report, 1967.

[101] Niewodniczanski, J. W., "Determination of Zinc and Lead in Ores by Portable X-ray Fluorescence Analyzer Employing ^{238}Pu Source," *Institute of Geological Science Radiogeology and Rare Minerals Unit*, London, Report No. 278, 1967.

[102] Ostrowski, K. W., Gorski, L., and Niewodniczanski, J. W. in ORNL-IIC-10, Baker, P. S. and Gerrard, M., eds., Oak Ridge National Laboratory, Oak Ridge, Tenn., Sept. 1967, pp. 746-763.

[103] Plotnikov, A. I. and Pshenickniy, G. A., "The Choice of Exciting Source for Avoiding Interferences and Diminishing the Mineralogic Effect in Nondispersive X-ray Analysis," *Problems of Exploration Geophysics,* Vol. 11, Leningrad, 1969, pp. 13-23 (in Russian).

[104] Pshenichniy, G. A., "On X-ray Radiometric Determination of Elements with Atomic Numbers Less Than 30 by Fluorescent to Scattered Gamma-Ray Ratio," *Problems of Exploration Geophysics,* Vol. 11, Leningrad, 1969, pp. 24-25 (in Russian).

[105] Preuss, L. E., Collins, H., Artman, C., and Bugenis, C., "Compilation of Beta-Excited X-Ray Spectra," TID.-22361, U.S. Atomic Energy Commission, 1966.

[106] Rhodes, J. R., Florkowski, T., and Cameron, J. F., "Analysis of Sulphur and Cobalt in Hydrocarbons," Research Group Report AERE R3925, United Kingdom Atomic Energy Authority, 1962.

[107] Rhodes, J. R., Ahier, Mrs. T. G., and Boyce, I. S., in *Radiochemical Methods of Analysis,* Vol. II, IAEA, Vienna, 1965, pp. 431-449.

[108] Rhodes, J. R., Packer, T. W., and Boyce, I. S., in *Radioisotope Instruments in Industry and Geophysics,* Vol. I, IAEA, Vienna, 1966, pp. 127-146.

[109] Rhodes, J. R., *The Analyst,* Vol. 91, 1966, pp. 683-699.

[110] Rhodes, J. R., in ORNL-IIC-10, Baker, P. S. and Gerrard, M., eds., Oak Ridge National Laboratory, Oak Ridge, Tenn., Sept. 1967, pp. 843-869.

[111] Rhodes, J. R., Furuta, T., and Hall, J. D., "X-ray Analysis Using Radioisotope Sources," Interim Technical Report No. 3, ORO-3224-12, Isotopes-Industrial Technology (TID-4500), July 1967.

[112] Rhodes, J. R., in ORNL-IIC-10, Baker, P. S. and Gerrard, M., eds., Oak Ridge National Laboratory, Oak Ridge, Tenn., Sept. 1967, pp. 442-464.

[113] Rhodes, J. R. and Furuta, T., in *Advances in X-ray Analysis,* Vol. 11, Newkirk, J. B., Mallett, G. R., and Pfeiffer, H. G., eds., Plenum Press, New York, 1968, pp. 249-274.

[114] Rhodes, J. R., "X-ray Analysis Using Radioisotope Sources," Final Report, ORO-3224-14, Isotopes-Industrial Technology (TID-4500), Sept. 1968.

[115] Rhodes, J. R. and Furuta, T., *Transactions of the Institution of Mining and Metallurgy,* Vol. 77, 1968, pp. B162-B165.

[116] Rhodes, J. R., Burton, B. S., and Furuta, T., "Feasibility of Nuclear Methods for the Determination of SO_2 in Stack Gases," TID-25137, 1968, U. S. Atomic Energy Commission, pp. 251-315.

[117] Rhodes, J. R., in *Non-Destructive Testing,* Egerton, H. B., ed., Oxford University Press, London, 1969, Chapter 14.

[118] Rhodes, J. R., Furuta, T., and Berry, P. F., in *Nuclear Techniques and Mineral Resources,* IAEA, Vienna, 1969 (STI/PUB/198), pp. 353-363.

[119] Rhodes, J. R., "Differential Ion Chambers," U.S. Patent 3,514,602, May 1970 (filed 16 Dec. 1966).

[120] Robert, A., Martinelli, P., Daniel, G., and Laflotte, J. L., "Radioisotopic Apparatus for Analyzing Low Atomic Number Elements by Fluorescence," Report CEA-N-1190, CEN, Saclay, 1969.

[121] Robinson, W. K., Adams, W. D., and Duggan, J. L., *American Journal of Physics,* Vol. 36, 1968, pp. 683-689.

[122] Sellers, B., Wilson, H. H., and Ziegler, C. A., ORO-3561-1, Division of Isotopes Development, U. S. Atomic Energy Commission, 1967.

[123] Sellers, B., Wilson, H. H., and Hanser, F. A., NYO-3491-3, Division of Isotopes Development, U. S. Atomic Energy Commission, 1968.

[124] Senftle, F. E. and Tanner, A. B., in ORNL-IIC-10, Baker, P. S. and Gerrard, M., eds., Oak Ridge National Laboratory, Oak Ridge, Tenn., Sept. 1967, pp. 503-521.

[125] Shmonin, L. I., Enker, M. B., Mager, E. V., and Oskapenko, V. F., "Use of Gamma-Excited X-ray Luminescence to Determine Lead Content in Multimetal Ores," CONF-651090-1, 1967.

[126] Smirnov, V. N., Ushkova, M. I., and Novikov, A. M., *Soviet Journal of Atomic Energy* (English Translation), Vol. 25, Vol. 5, 1968, p. 1235.

[127] Starnes, P. E. and Clarke, J. W. G., in *Radioisotope Instruments in Industry and Geophysics,* Vol. 1, IAEA, Vienna, 1966, pp. 243-269.

[128] Tolmie, R. W. in ORNL-IIC-10, Baker, P. S. and Gerrard, M., eds., Oak Ridge National Laboratory, Oak Ridge, Tenn., Sept. 1967, pp. 817-842.

[129] Uchida, K., Tominaga, H., Imamura, H., and Miwa, H., in *Radioisotope Instruments in Industry and Geophysics,* Vol. 1, IAEA, Vienna, 1966, pp. 113-126.

[130] Vertes, A. and Beck, I., *Journal of Radioanalytical Chemistry,* Vol. 2, 1969, pp. 323-334.

[131] Vertes, A. and Beck, I., *Journal of Radioanalytical Chemistry,* Vol. 2, 1969, pp. 335-343.

[132] Wasilewska, M. and Robert, A., "Dosage du Silicium dans un Silicate Soluble par une Methode Radioisotopique de Fluorescence X," CEA-R-3811, CEN, Saclay, 1969.

[133] Watt, J. S., *International Journal of Applied Radiation and Isotopes,* Vol. 18, 1967, pp. 383-391.

[134] Watt, J. S. in ORNL-IIC-10, Baker, P. S. and Gerrard, M., eds., Oak Ridge National Laboratory, Oak Ridge, Tenn., Sept. 1967, pp. 663-695.

[135] Watt, J. S., "Current and Potential Applications of Radioisotope X-ray and Neutron Techniques of Analysis in the Mineral Industry," *Annual Conference of the Australian Institution of Mining and Metallurgy,* Sydney, Australia, Aug. 1967, CONF-690815-5.

[136] White, G., "The Semi-continuous Analysis of Sinter Mixtures Using Radio-isotopes," *Nineteenth Chemists' Conference,* British Iron and Steel Research Association, 24 Buckingham Gate, London SW1, 1966, pp. 17-26.

[137] Yamamoto, S., *Analytical Chemistry,* Vol. 41, 1969, pp. 337-342.